Springer Theses

Recognizing Outstanding Ph.D. Research

Aims and Scope

The series "Springer Theses" brings together a selection of the very best Ph.D. theses from around the world and across the physical sciences. Nominated and endorsed by two recognized specialists, each published volume has been selected for its scientific excellence and the high impact of its contents for the pertinent field of research. For greater accessibility to non-specialists, the published versions include an extended introduction, as well as a foreword by the student's supervisor explaining the special relevance of the work for the field. As a whole, the series will provide a valuable resource both for newcomers to the research fields described, and for other scientists seeking detailed background information on special questions. Finally, it provides an accredited documentation of the valuable contributions made by today's younger generation of scientists.

Theses are accepted into the series by invited nomination only and must fulfill all of the following criteria

- They must be written in good English.
- The topic should fall within the confines of Chemistry, Physics, Earth Sciences, Engineering and related interdisciplinary fields such as Materials, Nanoscience, Chemical Engineering, Complex Systems and Biophysics.
- The work reported in the thesis must represent a significant scientific advance.
- If the thesis includes previously published material, permission to reproduce this must be gained from the respective copyright holder.
- They must have been examined and passed during the 12 months prior to nomination.
- Each thesis should include a foreword by the supervisor outlining the significance of its content.
- The theses should have a clearly defined structure including an introduction accessible to scientists not expert in that particular field.

More information about this series at http://www.springer.com/series/8790

Yuan Wang

Aerosol-Cloud Interactions from Urban, Regional, to Global Scales

Doctoral Thesis accepted by
Texas A&M University, College Station, USA

 Springer

Author
Dr. Yuan Wang
California Institute of Technology
Pasadena, CA
USA

Supervisor
Prof. Dr. Renyi Zhang
Texas A&M University
College Station, TX
USA

ISSN 2190-5053   ISSN 2190-5061   (electronic)
Springer Theses
ISBN 978-3-662-47174-6   ISBN 978-3-662-47175-3   (eBook)
DOI 10.1007/978-3-662-47175-3

Library of Congress Control Number: 2015938727

Springer Heidelberg New York Dordrecht London

Printed on acid-free paper

Springer-Verlag GmbH Berlin Heidelberg is part of Springer Science+Business Media
(www.springer.com)

Parts of this thesis have been published in the following journal articles:

Wang, Y., M. Wang, R. Zhang, S.J. Ghan, Y. Lin, J. Hu, B. Pan, M. Levy, J. Jiang, M.J. Molina, Assessing the Effects of Anthropogenic Aerosols on Pacific Storm Track Using A Multi-Scale Global Climate Model, *Proc. Natl. Acad. Sci. USA, 111* (19), 6894–6899 (2014)

Wang, Y., R. Zhang, R. Saravanan, Asian pollution climatically modulates mid-latitude cyclones following hierarchical modelling and observational analysis, *Nature. Comm., 5*, 3098 (2014)

Wang, Y., J. Fan, R. Zhang, R. Leung, C. Franklin, Improving Bulk Microphysics Parameterizations in Simulations of Aerosol Indirect Effects, *J. Geophys. Res., 118*, 1–19, (2013)

Fan, J., L. Leung, Z. Li, H. Morrison, H. Chen, Y. Zhou, Y. Qian, **Y. Wang**, Aerosol impacts on clouds and precipitation in eastern China-results from bin and bulk microphysics, *J. Geophys. Res., 117*, D00K36, (2012)

Wang, Y., Q. Wan, W. Meng, F. Liao, H. Tan, R. Zhang, Long-term impacts of aerosols on precipitation and lightning over the Pearl River Delta megacity area in China, *Atmo. Chem. Phys., 11*(23), 12421–12436, (2011)

Supervisor's Foreword

Yuan Wang received a B.S. in Computer Science from Fudan University (2007) and a Ph.D. (2013) in Atmospheric Sciences from Texas A&M University. He is presently a postdoc research fellow at the Jet Propulsion Laboratory, California Institute of Technology. His research areas focus on aerosol physics and chemistry, aerosol-cloud interactions and their climate implications, mesoscale and global climate modeling.

This dissertation covers several broad research topics in atmospheric sciences, ranging from development and refinement of the regional Weather Research Forecast (WRF) model and global climate model (GCM) to assessments of the aerosol-cloud-climate interaction. As recognized in the Fifth Assessment Report by the Inter-government Panel on Climate Change (IPCC AR5, 2013), the magnitude of cloud adjustment by aerosols is highly uncertain, representing the largest uncertainty in projections of future climate by anthropogenic activities. Much of the uncertainty in the assessment of the aerosol radiative forcing arises from the difficulty to resolve aerosol and cloud-scale processes in GCMs. Yuan's research involved modeling the aerosol effects on cloud formation and precipitation using the mesoscale cloud-resolving (CR) WRF model. He made key model development that, for the first time, implemented an explicit two-moment bulk microphysical scheme into WRF. Because of its broad application in numerical weather prediction, his improved WRF framework has enabled assessment of aerosol-cloud interaction for variable weather-scale systems, from individual cumulus to mesoscale convective systems. He utilized this model to investigate the aerosol effects on precipitation and lightning over a megacity region in China. He also improved the representation of cloud microphysics in the WRF model and employed the upgraded model to investigate the aerosol effects on stratocumulis over the Southeast Pacific region.

He developed a novel hierarchical modeling approach to assess the impacts of Asian pollution on the Pacific storm track, on the basis of the aerosol forcings calculated from seasonal simulations using the CR-WRF models and incorporation of the derived aerosol forcings to the GCM to predict the associated climatic responses. His novel approach resolved a long outstanding mismatch between simulations by regional mesoscale models and GCMs in the climate research community, allowing a more precise assessment of the global impacts of the cloud-scale processes induced by anthropogenic aerosols. His results from the hierarchical modeling and observational analysis demonstrated unambiguously that Asian pollution has exerted profound climatic impacts on mid-latitude cyclones over the past three decades. In addition, he evaluated the impacts of aerosols over the Pacific storm track on the basis of the results from multi-scale aerosol-climate model simulations, further confirming his results using the hierarchical modeling.

I believe that this dissertation will be of broad interest, not only to the scientific communities in the fields of atmospheric physics and chemistry, but also to those who are interested in global environmental and climate changes and the interface between science and policy.

College Station, USA
May 2015

Prof. Dr. Renyi Zhang
Editor, Journal of the Atmospheric Sciences
Senior Editor, Oxford Research
Encyclopedia—Environmental Science, Oxford University Press
Holder of Harold J. Haynes Endowed Chair
University Distinguished Professor
Department of Atmospheric Sciences, College of Geosciences
Department of Chemistry, College of Science
1204 O&M Building, 3150 TAMU
Texas A&M University
College Station, TX

Acknowledgments

I would like to express my deep gratitude to my thesis advisor, Dr. Renyi Zhang, for his wise guidance and unselfish support throughout the course of this research. My committee members, Dr. Ramalingam Saravanan, Dr. Courtney Schumacher, Dr. Qi Ying, and Dr. Yangang Liu are also highly appreciated for their valuable comments and constructive advises during the my doctoral study.

I want to thank my colleagues, the department faculty and staffs in the Department of Atmospheric Sciences at Texas A&M University. Especially, the discussions with Research Scientists Jenshan Hsieh and Cara-Lyn Lappen at Texas A&M University helped me solve several technical problems in numerical modeling. Special thanks go to Jiwen Fan from Pacific Northwest National Laboratory, who offered me substantial help during my visit to PNNL. I also benefited a lot from the collaboration with Minghuai Wang from PNNL to analyze the results from the super-parameterized climate model.

I also want to extend my gratitude to the National Aeronautics and Space Administration fellowship program, which provided me a 3-year funding support.

This thesis is specially dedicated to my beloved mother, Ruojing Zhang who passed away in 2011.

Contents

1 Introduction.   1
  1.1 Current Understanding on the Aerosol-Cloud Interaction.   1
  1.2 Urban-Scale Impacts of Air Pollution on the Thunderstorm   2
  1.3 Regional-Scale Impacts of Aerosol on the Stratocumuli.   3
  1.4 Global-Scale Impacts of Pollution Outflows on Storm Track   4
  1.5 Objectives.   5
  References   5

2 Numerical Model Description   9
  2.1 Mesoscale Weather Research and Forecast (WRF) Model   9
     2.1.1 Spectral Bin Cloud Microphysics.   9
     2.1.2 Bulk Cloud Microphysics—Morrison Scheme 10
     2.1.3 Bulk Cloud Microphysics in the CR-WRF 11
  2.2 Global Climate Model 13
  2.3 Multiscale Aerosol-Climate Modeling Framework 13
  References 14

3 Impacts of Urban Pollution on Thunderstorms 17
  3.1 Long-Term Observations of Precipitation,
        Lightning Flashes, and Visibility 17
  3.2 Design of Numerical Simulations. 20
  3.3 Model Evaluation and Sensitivities for Aerosol Levels 22
  3.4 Lightning Flashes and Lightning Potential Index 25
  3.5 Microphysical Properties and Convections. 28
  3.6 Summary 32
  References 34

**4 Aerosol Effects on the Stratocumulus and Evaluations
  of Microphysics** 37
  4.1 Experiment Design. 37

4.2   Effects of Aerosol Representation on Sc Simulations   39
       4.2.1   Simulated Aerosol Evolution   39
       4.2.2   Comparison with Field Measurements   40
       4.2.3   Effects on the Cloud Properties   43
4.3   Effects of Diffusional Growth Parameterizations.   45
4.4   Effects of Autoconversion Parameterizations   47
4.5   Effects of Aerosol Representation on AIE   49
4.6   Summary   50
References   52

5   Impacts of Asian Pollution Outflows on the Pacific
     Storm Track   55
     5.1   Observational Evidences   55
     5.2   A Hierarchical Modeling Approach   58
            5.2.1   Configuration of CR-WRF and Experiment Design   59
            5.2.2   Evaluations of CR-WRF Simulations   61
            5.2.3   Sensitivity Study and Derived Aerosol Forcings.   63
            5.2.4   Response of Storm Track in the Forced CAM5   70
     5.3   Multiscale Aerosol-Climate Modeling Framework   75
            5.3.1   Numerical Experiment Design   75
            5.3.2   Analysis of Simulation Results   75
            5.3.3   Results from Host GCM   80
     5.4   Summary   81
     References   82

6   Conclusions   85

List of Figures

Fig. 2.1   Overview of the two-moment bulk microphysical
            scheme (TAMU scheme) implemented in the WRF
            model.   12
Fig. 3.1   AOD and flash density distribution over Southern
            China and flash density distribution over Southern
            China. **a** Annual mean AOD from the MODIS satellite
            in 2005. **b** Annual mean CG flash density distribution
            from the local lightning detection network in the
            Guangdong Province in 2005 (Reprinted from Wang
            et al. (2011) with permission of Copernicus
            Publications)   18
Fig. 3.2   Correlation between **a** daily visibility and the heavy
            rainfall rate, **b** daily visibility and lightning flash
            density over PRD from 2000 to 2006. Both daily
            visibility and heavy rainfall rate are averaged over the
            4-month period (from March to June). The heavy
            rainfall rate is calculated from each rain gauge station
            with a daily rainfall greater than 25 mm. CG lightning
            flash density (flashes km^{-2}) is accumulated over
            4 months each year. The *line* represents a linear
            regression through all data (Reprinted from Wang et al.
            (2011) with permission of Copernicus Publications)   20
Fig. 3.3   Overview of domains in the model simulations. The
            red symbols represent the radar stations, and the *blue
            symbols* represent the lightning detection stations
            (Reprinted from Wang et al. (2011) with permission of
            Copernicus Publications).   21

Fig. 3.4   Comparison of radar reflectivity between observation
                and CR-WRF simulation for the P-case. **a** Observation
                at 0900 UTC. **b** Simulation at 0900 UTC.
                c Observation at 1100 UTC. **d** Simulation at 1100 UTC
                (Reprinted from Wang et al. (2011) with permission of
                Copernicus Publications).   23
Fig. 3.5   Temporal evolution of the domain-averaged rainfall
                rate. The *red dashed line* represents the gauge
                measurement, the *blue solid line* corresponds to the
                P-case, and the *green solid line* corresponds to the
                C-case (Reprinted from Wang et al. (2011) with
                permission of Copernicus Publications).   24
Fig. 3.6   Comparison of accumulated precipitation from 0900 to
                1500 UTC between observation and simulations.
                a Gauge measurements. **b** P-case. **c** C-case (Reprinted
                from Wang et al. (2011) with permission of Copernicus
                Publications)   24
Fig. 3.7   Probability distribution functions of the four different
                rainfall categories for the P- and C-cases. **a** Percentage of
                grid areas under a certain precipitation category over the
                entire model domain. **b** Percentage of the precipitation
                amount under a certain category over the total precip-
                itation amount. *Dark blue* corresponds to the P-case, and
                red corresponds to the C-case (Reprinted from Wang
                et al. (2011) with permission of Copernicus Publications)   26
Fig. 3.8   Temporal evolution of **a** observed total CG lightning
                flashes and **b** calculated domain-averaged LPI. In **a** *red*
                denotes positive flashes and *dark blue* denotes negative
                flashes. In **b** *red* denotes the C-case and *dark blue*
                denotes the P-case (Reprinted from Wang et al. (2011)
                with permission of Copernicus Publications)   27
Fig. 3.9   Comparison of observed CG lightning distribution
                with the simulated LPI in the P-case. **a** Strikes (0800
                UTC), **b** LPI (0800 UTC), **c** LPI (0800 UTC),
                d Strikes (0900 UTC), **e** LPI (0900 UTC), **f** LPI (0900
                UTC) (Reprinted from Wang et al. (2011) with
                permission of Copernicus Publications).   28
Fig. 3.10   Temporal evolution of the horizontally domain-sum-
                mated mass mixing ratio of **a** cloud water in the
                C-case, **b** cloud water in P-case, **c** rain water in the
                C-case, **d** rain water in the P-case, **e** ice in the C-case,
                f ice in the P-case, **g** graupel in the C-case, and
                h graupel in the P-case (Reprinted from Wang et al.
                (2011) with permission of Copernicus Publications)   30

Fig. 3.11   Domain averaged column maximum vertical velocity
              in the simulations. The *solid lines* represent the updraft
              and the *dashed lines* represent the downdraft. The *dark
              blue lines* represent the P-case and the *red lines*
              represent the C-case (Reprinted from Wang et al.
              (2011) with permission of Copernicus Publications)   32
Fig. 3.12   Temporal evolution of latent heat profiles: **a** C-case
              and **b** P-case (Reprinted from Wang et al. (2011) with
              permission of Copernicus Publications).   32
Fig. 4.1   **a** Domain overview for the simulation case. *Red line*
              denotes the flight track of C130 research aircraft on
              Oct 28, 2008. **b** Aerosol size distribution from field
              measurement and used in the model initialization
              (Reprinted from Wang et al. (2013) with permission of
              John Wiley and Sons)   38
Fig. 4.2   Temporal evolution of the domain-averaged aerosol
              number concentration from the three simulations with
              SBM, Bulk-OR, and Bulk-2M in the Sc case
              (Reprinted from Wang et al. (2013) with permission of
              John Wiley and Sons)   40
Fig. 4.3   Comparison of the vertical profiles of cloud
              microphysical properties from the three simulations
              SBM, Bulk-OR and Bulk-2M with the C130 aircraft
              measurements. The *first, second,* and *third columns*
              present the liquid water content, cloud droplet number
              concentration, and cloud droplet effective radius,
              respectively. The *black dots* denote the mean values of
              observations at given heights within ±25 m. *Shading
              areas* denote the standard derivation of the sampling
              data over the flight track (Reprinted from Wang et al.
              (2013) with permission of John Wiley and Sons).   41
Fig. 4.4   Comparison of time series of **a** LWP, **b** cloud base
              height, **c** cloud thickness, and **d** radar reflectivity at
              100 m from the three simulations SBM, Bulk-OR, and
              Bulk-2M with the C130 aircraft measurements. The
              error bar denotes the standard derivation of the
              sampling data over the flight track region at each
              altitude (Reprinted from Wang et al. (2013) with
              permission of John Wiley and Sons)   43

Fig. 4.5   Temporal evolution of the domain-averaged **a** cloud
            droplet number concentration, **b** cloud mass mixing
            ratio, **c** raindrop number concentration, **d** rain mass
            mixing ratio, **e** accumulated rainfall, and **f** core-area
            updraft velocity from the three simulations (Reprinted
            from Wang et al. (2013) with permission of John Wiley
            and Sons)   44
Fig. 4.6   Vertical profiles of cloud properties on cloud points
            from 0000 to 0100 UTC and from 0100 to 0300 UTC
            in the 3 h simulations using SBM and Bulk-2M
            without collision/sedimentation processes and radiation
            scheme (Reprinted from Wang et al. (2013) with
            permission of John Wiley and Sons)   46
Fig. 4.7   Temporal evolution of the domain-averaged cloud
            properties from four bulk simulations using the
            different autoconversion schemes and SBM (Reprinted
            from Wang et al. (2013) with permission of John Wiley
            and Sons)   48
Fig. 4.8   Comparisons of the domain-averaged cloud properties
            from simulations with SBM, Bulk-OR, and Bulk-2M
            under clean and polluted (control) aerosol conditions
            (Reprinted from Wang et al. (2013) with permission of
            John Wiley and Sons)   49
Fig. 5.1   Leading principal component of **a** 850 mb EMHF and
            b 300 mb EMWV over the northwest Pacific   57
Fig. 5.2   Spatial distribution of the EMHF at 850 hPa during
            a 1979–1988, **b** 2002–2011, **c** difference between the
            two decades and the EMWV at 300 hPa during
            d 1979–1988, **e** 2002–2011, **f** difference between the
            two decades over the northwest Pacific from NECP/
            DOE Reanalysis II Dataset. *Dots* in (**c**) and (**f**) indicate
            the 90 % level of statistical significance based on the
            Student's t-test.   58
Fig. 5.3   Spatial distribution of the eddy meridional heat flux
            (EMHF) at 850 hPa during **a** 1979–1988,
            b 2002–2011, **c** difference between the two decades
            and the eddy meridional wind variance (EMWV) at
            300 hPa during **d** 1979–1988, **e** 2002–2011,
            f difference between the two decades over the
            northwest Pacific from ECMWF ERA-Interim Dataset.
            Dots in (**c**) and (**f**) indicate the 90 % level of statistical
            significance based on the Student's t-test   59
Fig. 5.4   Domains overview for CR-WRF simulations   60

Fig. 5.5   Vertical profile of aerosol number concentration in the
               initial conditions of CR-WRF. **a** Number concentration
               profile of ammonium sulfate (cm^{-3}). *Red line* denotes
               M-case and *dark blue line* denotes P-case. **b** Number
               concentration profile of sea salt (cm^{-3}).   61
Fig. 5.6   Snapshot of storm event. **a** CR-WRF simulated cloud
               water path in Jan 20. **b** CR-WRF simulated cloud
               water path on Jan 21. **c** CR-WRF simulated cloud
               water path on Jan 22. **d** MODIS Terra L3 Daily cloud
               water path on Jan 20. **e** MODIS Terra L3 Daily cloud
               water path on Jan 21. **d** MODIS Terra L3 Daily cloud
               water path on Jan 22 (Reprinted from Wang et al.
               (2014a) with permission of Nature Publication Group)   62
Fig. 5.7   Comparison of cloud fraction. **a** MODIS L3 monthly
               cloud fraction in January. **b** WRF simulated cloud
               fraction in January. **c** MODIS L3 monthly cloud
               fraction in February. **d** WRF simulated cloud fraction
               in February (Reprinted from Wang et al. (2014a) with
               permission of Nature Publication Group)   64
Fig. 5.8   Comparison of precipitation. **a** TRMM monthly
               accumulated rainfall in January. **b** WRF simulated
               monthly accumulated rainfall in January. **c** TRMM
               monthly accumulated rainfall in February. **d** WRF
               simulated monthly accumulated rainfall in February   65
Fig. 5.9   Temporal evolution of domain-averaged TOA
               a shortwave forcing. **b** longwave forcing and **c** net
               forcing. 5-day smoothing is employed. *Dark blue lines*
               represent P-case and *red lines* represent M-case.   66
Fig. 5.10   Spatial distribution of 2-month averaged daily
               precipitation **a** in P-case and **b** in M-case. **c** Temporal
               evolution of domain-averaged precipitation rate. 5-day
               smoothing is employed. *Dark blue lines* represent
               P-case and *red lines* represent M-case   67
Fig. 5.11   Time series of domain-averaged **a** cloud number
               concentration, **b** cloud effective radius, **c** cloud water
               path, **d** cloud ice path, **e** cloud optical thickness and
               f cloud core area percentage. 5-day smoothing is
               employed. *Dark blue lines* represent P-case and *red
               lines* represent M-case   68

Fig. 5.12   Heating rate profiles from CR-WRF simulations. The
             blue lines denote the heating rates from P-case and red
             lines for M-case. The *black dot-dash lines* denote the
             heating difference between P-case and M-case. In (**e**),
             the *dash lines* denote the latent heat release rates
             (positive only) and the *dot-dot-dash lines* denote the
             cooling (negative only) rates   70
Fig. 5.13   Location of additional heating in the GCM domain   71
Fig. 5.14   Comparison of EMHF and EMWV between CTRL
             and AERO   72
Fig. 5.15   Comparison of temperature profiles between CTRL
             and AERO. *Dark blue line* denotes the profile from
             AERO, *red line* denotes the profile from CTRL, and
             the *dash line* denotes the difference between AERO
             and CTRL   73
Fig. 5.16   Comparison of maximum updraft velocities between
             CTRL and AERO. The *blue line* denotes the result
             from AERO and the *red line* denotes that from CTRL   73
Fig. 5.17   Comparison of high cloud fractions between CTRL
             and AERO (Reprinted from Wang et al. (2014a) with
             permission of Nature Publication Group)   74
Fig. 5.18   **a** The difference of aerosol optical depth (AOD)
             between PD and PI over NW Pacific. **b** The
             comparison of aerosol mass concentration and
             chemical composition in the accumulation mode
             between PD and PI over the NW Pacific in
             PNNL-MMF (Reprinted from Wang et al. (2014b)
             with permission of the National Academy of Sciences
             of the United States of America)   76
Fig. 5.19   The difference of **a** cloud number concentration (Nc),
             b liquid water path (LWP), **c** ice water path (IWP),
             d high cloud fraction, **e** shortwave cloud radiative
             forcing (SWCF), **f** longwave cloud radiative forcing
             (LWCF), **g** precipitation (PREC) and **h** eddy
             meridional heat flux at 850 mb (EMHF) between PD
             and PI over the NW Pacific in PNNL-MMF. *Black dots*
             indicate the regions with significance of t-test larger
             than 90 % (Reprinted from Wang et al. (2014b) with
             permission of the National Academy of Sciences of the
             United States of America)   77

Fig. 5.20   The difference of vertical distribution of **a** convective
             cloud fraction and **b** cloud top fraction between PD and
             PI over the NW Pacific in PNNL-MMF (Reprinted
             from Wang et al. (2014b) with permission of the
             National Academy of Sciences of the United States of
             America)   79
Fig. 5.21   The difference of **a** aerosol optical depth (AOD),
             b liquid water path (LWP), **c** ice water path (IWP) and
             d precipitation (PREC) between PD and PI over the
             NW Pacific in CAM5 (Reprinted from Wang et al.
             (2014b) with permission of the National Academy of
             Sciences of the United States of America).   80

List of Table

Table 3.1   Domain-averaged properties of hydrometeors in the
CR-WRF simulations   29

Chapter 1
Introduction

1.1 Current Understanding on the Aerosol-Cloud Interaction

Produced from anthropogenic and natural sources (IPCC 2007; Zhang et al. 2004, 2009), atmospheric aerosols play an important role in regulating the radiative balance of the Earth-Atmosphere system, directly by reflecting or absorbing the incoming solar radiation and indirectly by influencing cloud formation. In particular, aerosols by serving as cloud condensation nuclei (CCN) or ice nuclei (IN), also known as the aerosol indirect effect (AIE), may considerably impact the lifetime, albedo, and precipitation of cloud systems, through a complex interaction between cloud microphysics and dynamics (Williams 2002; van den Heever and Cotton 2007). Therefore, atmospheric aerosols have been closely linked with modification of cloud systems of diverse scales, ranging from isolated convective storms (Fan et al. 2007; Li et al. 2008b), mesoscale convective systems such as squall lines (Li et al. 2009; Tao et al. 2007) and hurricanes (Khain et al. 2008), to large-scale circulations such as wintertime Pacific storm track (Zhang et al. 2007; Li et al. 2008a) and summertime Asian monsoon (Lau et al. 2006). Currently, the direct and indirect forcings of aerosols on climate are highly uncertain, representing the largest uncertainty in climate predictions ($+0.8/-1.5$ Wm^{-2}) (IPCC 2007). For example, the aerosol effects on different types of cloud and precipitation systems have been shown to be highly nonlinear, which poses large challenges on quantification of the aerosol indirect forcing (Li et al. 2008b; Khain 2009; Ntelekos et al. 2009; Tao et al. 2012).

Substantial efforts have been made to examine aerosol-cloud interaction under different atmospheric conditions using satellite and in situ measurements. Li et al. (2011) analyzed the 10-year dataset of aerosol, cloud and meteorological variables collected in the Southern Great Plains in the United States and found out that cloud-top height and thickness increase with aerosol concentration measured near the ground in mixed-phase clouds but exhibit no sensitivity in the clouds with no ice or

© Springer-Verlag Berlin Heidelberg 2015
Y. Wang, *Aerosol-Cloud Interactions from Urban, Regional, to Global Scales*,
Springer Theses, DOI 10.1007/978-3-662-47175-3_1

cool bases. Koren et al. (2012) shows that in the tropics and subtropics, increases in aerosol abundance observed by Moderate-Resolution Imaging Spectroradiometer (MODIS) are associated with the local intensification of rain rates detected by the Tropical Rainfall Measuring Mission (TRMM).

Several mechanisms have been proposed from the modeling community to elaborate the role of aerosols on cloud development and precipitation under different dynamic and thermodynamic scenarios. It has been suggested that the aerosol effects on cloud properties vary with convective available potential energy (CAPE) and wind shear (Lee et al. 2008) and depend on the morphology of clouds (Lee et al. 2010). Fan et al. (2007) indicated that relative humidity plays an important role in the regulating aerosol effects. Another recent study by Fan et al. (2009) revealed a dominant role of wind shear on aerosol-cloud interaction. In addition, cold pool produced by evaporative cooling has been demonstrated to considerably modulate the influence of aerosols on the development of convective systems (van den Heever and Cotton 2007; Tao et al. 2007). Khain (2009) further suggested that heat and condensate mass budgets need to be evaluated when considering the effect of aerosols on precipitation. In addition, it has been suggested that the cloud droplet effective radius may increase or decrease with aerosol loading, depending on cloud dynamic and thermodynamic conditions and aerosol properties (Yuan et al. 2008).

Overall, current understanding of aerosol impacts on the radiative budget and hydrological cycle of the climate system is still inadequate at the fundamental level. Hence it is valuable and necessary to illustrate the aerosol indirect effect through case studies at different scales. In my research, I identify three regimes of interests to carry on the investigation of aerosol-cloud-climate interaction.

1.2 Urban-Scale Impacts of Air Pollution on the Thunderstorm

With fast economic and industrial development in the Eastern and Southern Asian countries, elevated anthropogenic pollutants from coal and biomass burning, industry emissions, etc., have caused severe episodes of air pollution around megacities. In the region of southern China called the Pearl River Delta (PRD) area, the occurrence of low visibility (<10 km) has remained highly frequent (150 days per year) since 1980, on the basis of the long-term visibility observation (Deng et al. 2008). Aerosol optical depth (AOD) retrieved from the Moderate-resolution Imaging Spectroradiometer (MODIS) satellite shows a typical value of larger than 0.6 in the PRD region, because of the existence of a persistent haze layer, which is also referred to as the Asian Brown Cloud (Wu et al. 2005). In addition to primary emissions of natural and anthropogenic origins, photochemical oxidation of anthropogenic and biogenic inorganic and organic compounds leads to nucleation and growth of secondary aerosols, contributing to particulate matter pollution in this region (Zhang et al. 2004; Wang et al. 2010). Several major field campaigns of air quality studies, such as the Program of Regional Integrated Experiments of Air Quality in the Pearl River Delta (PRIDE-PRD) 2004 and 2008 (Zhang et al. 2008), have been conducted in this region to monitor the pollution situation and characterize the chemical and meteorological conditions responsible for accumulations of gaseous and PM pollutants of primary and secondary origins.

In the metropolitan areas, the plausible effect of urban aerosols on lightning enhancement in thunderstorms was first suggested by Westcott (1995). Orville et al. (2001) examined cloud-to-ground lightning flashes from the National Lightning Detection Network (NLDN) in Houston, Texas for a 12-year period and found higher lightning flash density near the urban area. Over major urban areas of South Korea, enhancement of cloud-to-ground lightning under high aerosols loading was documented on the basis of measured lightning flashes and particulate matter (PM) (Kar et al. 2009). According to the long-term ground lightning observation and results from the Lightning Imaging Sensor (LIS) database, the Pearl River Delta (PRD) area ($113°–114.5°$ E, $21.5°–23.5°$ N) with a cluster of large cities including, Guangzhou, Shenzhen, Hong Kong and Macau, exhibits frequent lightning strikes and has a lightning density of 31.4 flashes km^{-2} $year^{-1}$, the largest value observed in China (Ma 2005). The thermodynamic and dynamical conditions in this area are favorable for cloud electrification and lightning formation, since the region is located in the subtropical and coastal area with abundant solar heating and moisture sources. In addition, urbanization in this area forms a cluster of large cities with urban land surface characteristics such as increased roughness, decreased moisture availability, and decreased thermal inertia, leading to a prominent urban heat island effect. Therefore, it is valuable to investigate the effects of coupling between the urban pollution and the unstable atmospheric condition over the PRD area.

1.3 Regional-Scale Impacts of Aerosol on the Stratocumuli

Maritime stratocumuli (Sc) represent one of the most frequent and important categories of cloud systems over the globe. Realistic simulation of this cloud system is crucial for assessment of cloud radiative forcing induced by aerosols in the global radiative budget, considering several important characteristics of maritime stratocumulus, such as persistent cloud cover, high reflectivity, and low cloud-base height. Over the southeast Pacific, various components are coupled in the southeast Pacific climate system, including the large-scale subsidence, orographic effect from Andes Mountains, and the coastal currents and upwelling. All of them coordinately contribute to the large area of persistent stratocumulus deck in the maritime boundary layer. The occasional outflows of continental aerosols impose perturbation to the cloud lifetime, cloud albedo, and drizzle precipitation associated with the stratocumulus deck. Recent studies suggest that perturbations of cloud structure and energy balance from increased anthropogenic emissions are large in the susceptible clean environment (Wang et al. 2010; Yang et al. 2011). In 2008, there was a field campaign VAMOS Ocean-Cloud-Atmosphere-Land Study (VOCALS) carried out over the southeast Pacific with the objective to improve the scientific understanding of interactions between different components in regional climate systems. Abundant in situ measurements of aerosols and clouds from various research aircraft and ships largely facilitate our investigation of aerosol impacts on the stratocumulus cloud in this region.

The characteristic dynamical processes and relatively simple microphysical associated with the warm Sc system is also favorable to evaluate cloud microphysical parameterizations in numerical models. Different aerosol representation approaches, different treatments of diffusional growth of cloud droplets, and different autoconversion schemes to form raindrop will be investigated in this cloud regime.

1.4 Global-Scale Impacts of Pollution Outflows on Storm Track

Over the north Pacific region, atmospheric conditions such as cloudness, storminess, and precipitation have significantly changed during the past few decades. Graham and Diaz (2001) analyzed the NCEP–NCAR reanalysis and in situ data and found that the frequency and intensity of extreme cyclones have increased significantly in the North Pacific Ocean over the past 50 years. Zhang et al. (2007) suggested that the amount of deep convective cloud associated with the Pacific storm track increased in wintertime over the north Pacific from 1984 to 2005 through analysis of the cloud satellite data from International Satellite Cloud Climatology Project (ISCCP) and high-resolution infrared sounder (HIRS) satellites. As a signature of the storm track, precipitation over the north Pacific has been studied by Li et al. (2008a) using data from Global Precipitation Climatology Project (GPCP), and it is reported that precipitation over the North Pacific increases by about 1.5 mm/year in winter during the period of 1984–2005.

Meanwhile, there were tremendous pollutions produced in South and East Asian countries along with their fast economic growth during a few past decades. Measurements from field campaigns, such as the Transport and Chemical Evolution over the Pacific (TRACE-P) and Asian Pacific Regional Aerosol Characterization Experiment (ACE-Asia), revealed the intensive Asian continental aerosol outflow to the west Pacific (Jacob et al. 2003; Huebert et al. 2008) and even to the North American continent (Yu et al. 2012). Visible images from NASA satellite clearly show transportation of the haze plume over from the East Asia to the Pacific. Transportation of pollution plume can also be identified through the MODIS satellite measurement, which estimated that about 18 Tg/year pollution aerosol was exported from east Asia to the northwestern Pacific Ocean (Yu et al. 2008).

Numerical studies based on regional climate model have examined the direct radiative effect of aerosols on the climate of the storm track region (Gong et al. 2006; Fischer-Bruns et al. 2009). Cloud resolving models have been utilized to investigate the aerosol indirect effect on the cloud properties of a single storm through short-term simulations (Zhang et al. 2007; Li et al. 2008a). In my research, the different modeling approaches are used to quantify the impacts of Asian pollution on the wintertime Pacific storm track with the focus on the influences of the large-scale dynamics.

1.5 Objectives

The primary objective in the present study is to advance our understanding of aerosol-cloud-climate interaction in different atmospheric conditions with different spatial scales. The specific tasks of this study includes: (1) perform an analysis of 7-year measurements of precipitation, lightning flashes, and visibility from 2000 to 2006 are analyzed over the PRD area, China; (2) employ Weather Research and Forecasting (WRF) model with a two-moment bulk microphysical scheme to simulate the aerosol impacts on the precipitation and lightning of a thunderstorm event in this area; (3) implement a two-moment prognostic aerosol approach in the Morrison microphysics scheme of WRF model, test some of cloud microphysical parameterizations in the warm-cloud physics, and conduct sensitivity study of aerosol effects on the stratocumulus clouds over Southeast Pacific; (4) derive and identify the interannual trend of NW Pacific storm track intensity based on reanalysis data; (5) conduct long-term WRF simulation over the North Pacific to examine the responses of storm track under different continental aerosol loadings; (6) implement the aerosol forcings derived from CR-WRF to the Community Atmosphere Model Version 5.1 (CAM5) and performed the long-term simulation to assess the climate effect of the modulation on Pacific storm track induced by Asian pollution; (7) assess the aerosol impacts on the Pacific storm track using the results from multi-scale aerosol-climate model simulations.

References

Deng X, Tie X, Wu D, Zhou X, Bi X, Tan H, Li F, Jiang C (2008) Long-term trend of visibility and its characterizations in the Pearl River Delta (PRD) region, China. Atmos Environ 42 (7):1424–1435

Fan J, Zhang R, Li G, Tao W-K (2007) Effects of aerosols and relative humidity on cumulus clouds. J Geophys Res 112(D14):D14204

Fan J, Yuan T, Comstock JM, Ghan S, Khain A, Leung LR, Li Z, Martins VJ, Ovchinnikov M (2009) Dominant role by vertical wind shear in regulating aerosol effects on deep convective clouds. J Geophys Res 114(D22):D22206

Fischer-Bruns I, Banse DF, Feichter J (2009) Future impact of anthropogenic sulfate aerosol on North Atlantic climate. Clim Dyn 32:511–524

Gong SL, Zhang XY, Zhao TL, Zhang XB, Barrie LA, McKendry IG, Zhao CS (2006) A simulated climatology of Asian dust aerosol and its trans-Pacific transport. Part II: Interannual variability and climate connections. J Clim 19:104–122

Graham NE, Diaz HF (2001) Evidence for intensification of North Pacific winter cyclones since 1948. Bull Am Meteor Soc 82:1869–1893

Huebert BJ, Bates T, Russell PB, Shi G, Kim YJ, Kawamura K, Carmichael G, Nakajima T (2008) An overview of ACE-Asia: strategies for quantifying the relationships between Asian aerosols and their climatic impacts. J Geophys Res 108(D23):8633

IPCC (2007) Climate change 2007: the physical science basis. In: Solomon S, Qin D, Manning M, Chen Z, Marquis M, Averyt KB, Tignor M, Miller HL (eds) Contribution of working group I to the fourth assessment report of the intergovernmental panel on climate change. Cambridge University Press, Cambridge

Jacob DJ et al (2003) Transport and chemical evolution over the Pacific (TRACE-P) aircraft mission: design, execution, and first results. J Geophys Res 108(D20):9000

Kar S, Liou Y, Ha K (2009) Aerosol effects on the enhancement of cloud-to-ground lightning over major urban areas of South Korea. Atmos Res 92(1):80–87

Khain AP (2009) Notes on state-of-the-art investigations of aerosol effects on precipitation: a critical review. Environ Res Lett 4(1):015004

Khain A, Cohen N, Lynn B, Pokrovsky A (2008) Possible aerosol effects on lightning activity and structure of hurricanes. J Atmos Sci 65(12):3652–3677

Koren I, Altaratz O, Remer LA, Feingold G, Martins JV, Heiblum RH (2012) Aerosol-induced intensification of rain from the tropics to the mid-latitudes. Nat Geosci 5(2):118–122

Lau KM, Kim MK, Kim KM (2006) Asian summer monsoon anomalies induced by aerosol direct forcing: the role of the Tibetan Plateau. Clim Dyn 26(7–8):855–864

Lee SS, Donner LJ, Phillips VTJ, Ming Y (2008) The dependence of aerosol effects on clouds and precipitation on cloud-system organization, shear and stability. J Geophys Res 113(D16): D16202

Lee SS, Donner LJ, Penner JE (2010) Thunderstorm and stratocumulus: how does their contrasting morphology affect their interactions with aerosols? Atmos Chem Phys 10(14):6819–6837

Li G, Wang Y, Lee K-H, Diao Y, Zhang R (2008a) Increased winter precipitation over the North Pacific from 1984–1994 to 1995–2005 inferred from the Global Precipitation Climatology Project. Geophys Res Lett 35(13). doi:10.1029/2008GL034668

Li G, Wang Y, Zhang R (2008b) Implementation of a two-moment bulk microphysics scheme to the WRF model to investigate aerosol-cloud interaction. J Geophys Res 113(D15):D15211

Li G, Wang Y, Lee K-H, Diao Y, Zhang R (2009) Impacts of aerosols on the development and precipitation of a mesoscale squall line. J Geophys Res 114(D17):D17205

Li ZQ, Niu F, Fan JW, Liu YG, Rosenfeld D, Ding YN (2011) Long-term impacts of aerosols on the vertical development of clouds and precipitation. Nat Geosci 4(12):888–894

Ma M (2005) Climatological distribution of lightning density observed by satellites in China and its circumjacent regions. Sci China Ser D 48(2):219

Ntelekos AA, Smith JA, Donner L, Fast JD, Gustafson WI, Chapman EG, Krajewski WF (2009) The effects of aerosols on intense convective precipitation in the northeastern United States. Q J R Meteorol Soc 135(643):1367–1391

Orville RE, Huffines G, Nielsen-Gammon J, Zhang R, Ely B, Steiger SM, Phillips S, Allen S, Read W (2001) Enhancement of cloud-to-ground lightning over Houston, Texas. Geophys Res Lett 28(13):2597–2600

Tao W-K, Li X, Khain A, Matsui T, Lang S, Simpson J (2007) Role of atmospheric aerosol concentration on deep convective precipitation: cloud-resolving model simulations. J Geophys Res 112(D24):D24S18

Tao W-K, Chen J-P, Li Z, Wang C, Zhang C (2012) Impact of aerosols on convective clouds and precipitation. Rev Geophys 50(2). doi:10.1029/2011RG000369

van den Heever SC, Cotton WR (2007) Urban aerosol impacts on downwind convective storms. J Appl Meteorol Climatol 46(6):828–850

Wang H, Feingold G, Wood R, Kazil J (2010) Modelling microphysical and meteorological controls on precipitation and cloud cellular structures in Southeast Pacific stratocumulus. Atmos Chem Phys 10(13):6347–6362

Westcott NE (1995) Summertime cloud-to-ground lightning activity around major midwestern urban areas. J Appl Meteorol 34:1633–1642

Williams E (2002) Contrasting convective regimes over the Amazon: implications for cloud electrification. J Geophys Res 107(D20):8082

Wu D, Tie X, Li C, Ying Z, Kai-Hon Lau A, Huang J, Deng X, Bi X (2005) An extremely low visibility event over the Guangzhou region: a case study. Atmos Environ 39(35):6568–6577

Yang Q, Gustafson JWI, Fast JD, Wang H, Easter RC, Morrison H, Lee YN, Chapman EG, Spak SN, Mena-Carrasco MA (2011) Assessing regional scale predictions of aerosols, marine stratocumulus, and their interactions during VOCALS-REx using WRF-Chem. Atmos Chem Phys 11(23):11951–11975

Yu HB, Remer LA, Chin M, Bian HS, Kleidman RG, Diehl T (2008) A satellite-based assessment of transpacific transport of pollution aerosol. J Geophys Res 113(D14). doi:10.1029/2007JD009349

Yu HB, Remer LA, Chin M, Bian HS, Tan Q, Yuan TL, Zhang Y (2012) Aerosols from overseas rival domestic emissions over North America. Science 337(6094):566–569

Yuan T, Li Z, Zhang R, Fan J (2008) Increase of cloud droplet size with aerosol optical depth: an observation and modeling study. J Geophys Res 113:D04201. doi:10.1029/2007JD008632

Zhang R, Suh I, Zhao J, Zhang D, Fortner EC, Tie X, Molina LT, Molina MJ (2004) Atmospheric new particle formation enhanced by organic acids. Sciences 304:1487–1490

Zhang R, Li G, Fan J, Wu DL, Molina MJ (2007) Intensification of Pacific storm track linked to Asian pollution. Proc Natl Acad Sci USA 104(13):5295–5299

Zhang YH, Hu M, Zhong LJ, Wiedensohler A, Liu SC, Andreae MO, Wang W, Fan SJ (2008) Regional integrated experiments on air quality over Pearl River Delta 2004 (PRIDE-PRD2004): overview. Atmos Environ 42(25):6157–6173

Zhang R, Wang L, Khalizov AF, Zhao J, Zheng J, McGraw RL, Molina LT (2009) Formation of nanoparticles of blue haze enhanced by anthropogenic pollution. Proc Natl Acad Sci USA 106:17650–17654. doi:10.1073/pnas.0910125106

Chapter 2
Numerical Model Description

Abstract To deal with the cloud systems in the diverse scales, various numerical simulation tools are used in this study. Two distinct modeling frameworks, cloud-resolving model and general circulation model, are employed for different cases on the basis of their own features and limitations. The combination of those two frameworks is also explored and evaluated in this study.

2.1 Mesoscale Weather Research and Forecast (WRF) Model

The Weather Research and Forecasting (WRF) Model is a next-generation meso-scale numerical weather prediction system designed to serve both atmospheric research and operational forecasting needs. It has a fully compressible Euler non-hydrostatic dynamics core. Arakawa C-grid staggering is used for horizontal grid, and vertical grid stretching is permitted. The representation of the microphysical processes in WRF model is essential in determining the cloud structure and development. Two types of microphysical schemes are commonly adopted to describe the size dependence of particles in cloud-resolving models, i.e., the spectral bin microphysics and bulk microphysics.

2.1.1 Spectral Bin Cloud Microphysics

With the utilization of several tens of bins to describe the number and mass distributions of hydrometeors and aerosols, spectral bin microphysics explicitly represents the physical processes on the cloud-resolving scale. A fast version of bin microphysics scheme based on Khain et al. (2004) (hereafter referred to as SBM) has been incorporated into the WRF model (Lynn et al. 2005; Khain et al. 2009; Fan et al. 2012). The fast version of SBM retains the advantages of the full SBM in

© Springer-Verlag Berlin Heidelberg 2015
Y. Wang, *Aerosol-Cloud Interactions from Urban, Regional, to Global Scales*,
Springer Theses, DOI 10.1007/978-3-662-47175-3_2

Khain et al. (2005) and produces cloud microphysical and dynamical structure as well as precipitation similar to the full SBM (Khain et al. 2009). SBM uses four size distribution spectra to represent hydrometeors and CCN in the model, including water drops (cloud and rain), low-density ice (ice and snow), high-density ice (graupel and hail), as well as CCN. Each spectrum is composed of 33 mass bins and the relationship between adjacent bins is determined by the function $m_k = 2^* \ m_{k-1}$.

In the SBM, supersaturation is explicitly predicted at each time step, and the critical radius of CCN (r_{Ncrit}) is calculated according to the Köhler theory using the value of supersaturation. At each time step, CCN with radius greater than r_{Ncrit} is removed from the CCN spectrum and the mass of the activated droplets is added to the cloud spectrum. The size of the activated cloud droplet is calculated under the assumption of equilibrium over the activated droplets, if the radius of the original CCN particle (r_N) is less than 0.03 μm. For large CCN particles with radius greater than 0.03 μm, the mass of water condensation on the particle is parameterized as $m_w = K(4/3)\pi r_N^3 \rho_w$, where $K = 5$ is used in this study (Khain et al. 2000).

Since supersaturation is explicitly predicted in the SBM by solving the equation for supersaturations with respect to water and ice, the diffusion growth/evaporation rate of liquid drops is directly calculated based on the supersaturation in each grid cell. The associated numerical equations are taken from Pruppacher and Klett (1997) and fully discussed by Khain et al. (2005). To better resolve the condensation/evaporation processes in the SBM, sub-timestep iteration is employed and the condensation/evaporation rate is calculated over each sub-time step Δt_{diff}. In the SBM simulation, we set $\Delta t = 4^* \Delta t_{diff}$. To avoid artificial broadening in the droplet spectrum as a result of diffusional growth and collisions, the remapping scheme are updated to conserve three moments of the hydrometeor size distributions (i.e., concentration, mass, and radar reflectivity), in contrast to the commonly used scheme of Kovetz and Olund (1969) that conserves only concentration and mass during the remapping (Khain et al. 2008). However, because of large computational resources demanded by the bin microphysics, the bulk microphysics represents a more practical choice for long-term simulations in regional and global climate models.

2.1.2 Bulk Cloud Microphysics—Morrison Scheme

By definition, the size distribution of the cloud hydrometeors in bulk microphysics is determined by the bulk properties of hydrometeors, such as number concentration and mass mixing ratio. The shape of each hydrometeor spectrum is prescribed by a certain type of distribution function, such as the Gamma or Marshall–Palmer function. Empirical parameterizations are needed in the bulk microphysics to resolve subgrid processes in relatively coarse resolution regional and global scale simulations.

A two-moment bulk microphysics with prognostic cloud number concentrations developed by Morrison and co-authors (Morrison et al. 2005, 2007; Solomon et al.

2009) is presently available for application under the framework of WRF (hereafter referred to as 'Bulk-OR'). There are two options in Bulk-OR to calculate the aerosol activation rate. One simple treatment is to parameterize the fraction of activated aerosols by the updraft velocity and assume the simple power-law CCN spectra using two empirical parameters (Fan et al. 2012). A more sophisticated treatment considers a wider range of governing parameters (Abdul-Razzak et al. 1998). In the latter parameterization, each mode of the aerosol spectrum is represented by a single lognormal size distribution. In Bulk-OR, all three parameters (the geometric mean radius, the total number concentration of aerosols, and the geometric standard deviation) associated with the lognormal size distribution are prescribed for each distribution function. There is no degree of freedom in the aerosol spectra, and the amount of aerosols that serve as CCN to form cloud droplets remains invariant in the simulation.

To overcome this deficiency, two prognostic variables of aerosols, i.e., the aerosol number concentration and mass mixing ratio, are introduced into the bulk scheme in the present work (hereafter referred to as Bulk-2M). During the activation process, activated aerosols to form cloud droplets are removed from the aerosol spectra and added to the cloud droplet spectra. The aerosol number and mass concentration are fixed at the lateral boundaries of the domain, to represent the external source of aerosols. By predicting both the number and mass concentration of aerosols, the total aerosol amount and the geometric mean radius of the aerosol spectra are time-dependent, because of removal of aerosols to form cloud droplets or addition of aerosols from the boundary sources. Advective and convective transports of aerosols are accomplished by the dynamical core of the model. The regeneration of aerosols due to evaporation of cloud droplets and raindrops is not included in the current scheme. The fraction of activated aerosols at each time step follows the parameterization developed by Abdul-Razzak et al. (1998).

In the Morrison scheme, the supersaturation is not calculated explicitly and a simplified liquid saturation adjustment strategy is utilized following the equation from Dudhia (1989). The default autoconversion scheme used in the Morrison scheme is a two-moment parameterization developed by Khairoutdinov and Kogan (2000) (hereafter referred to as KK2000). The cloudy grid cells are maintained at 100 % relative humidity. If the air is sub-saturated, cloud/rain water is evaporated continuously until saturation is reached within a time step, or if the air is supersaturated, vapor condensation removes the supersaturation within that time step. The autoconversion rate is derived by fitting the results of large eddy simulations of marine boundary layer clouds that used explicit bin microphysics with analytical functions.

2.1.3 Bulk Cloud Microphysics in the CR-WRF

Li et al. (2008, 2009) has implemented a two-moment bulk microphysical scheme with a three-moment aerosol approach into the WRF model (hereafter referred as

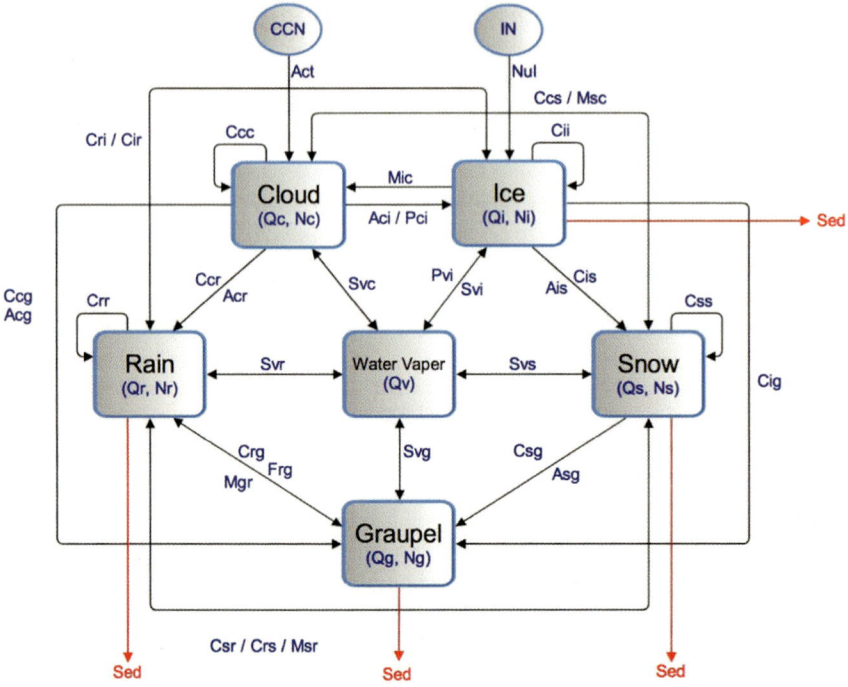

Fig. 2.1 Overview of the two-moment bulk microphysical scheme (TAMU scheme) implemented in the WRF model

CR-WRF). A description of the bulk microphysical processes in CR-WRF pertinent to this present work is summarized in the Fig. 2.1. Briefly, the number concentration and mass mixing ratio of five hydrometeors, including cloud, rain, ice, snow, and graupel, are explicitly predicted in the scheme with the assumption of gamma size distribution. Thirty-four microphysical processes are considered in the scheme, including seven autoconversion parameterization options for the warm rain processes and three types of heterogeneous ice nucleation. In contrast to treating the number concentration of cloud droplets or cloud condensation nuclei (CCN) to be constant in many other two-moment bulk microphysical schemes, three new prognostic variables, i.e., number concentration, surface area, and mass mixing ratio of aerosols, are implemented in this scheme. The process of aerosol activation into cloud droplet is modeled using the classic Köhler theory and the cloud number concentration is directly predicted from aerosol concentration and ambient relative humidity. Under the assumption of a log-normal size distribution of aerosols, the model approach with 48 bins is used to depict the aerosol spectrum in the scheme. The performances of this microphysical scheme under different conditions have been evaluated and an inter-comparison with other available microphysical schemes in the WRF V2.2.1 has been reported previously (Li et al. 2008, 2009).

One important feature of the microphysics in CR-WRF is to employ the new Kessler-type autoconversion scheme derived by Liu and Daum (2004) and Liu et al. (2004), which suggests a strong dependence of the autoconversion rate on the relative dispersion of the cloud droplet size distribution in addition to liquid water content and droplet concentration. Therefore, by incorporating the relative dispersion of the cloud droplet size distribution in this scheme, the coarse assumption inherent in the traditional Kessler-type parameterizations, such as a fixed collision kernel, can be eliminated.

2.2 Global Climate Model

Mesoscale cloud-resolving models with explicit cloud microphysics and aerosol representation are able to perform physically realistic simulations but unable to provide feedbacks to the large-scale circulation. Global climate models (GCM) are commonly used to investigate the responses of large-scale systems from different atmospheric forcings. The Community Earth System Model (CESM) developed by the National Center of Atmospheric Research (NCAR) is a coupled climate model with five separate models simultaneously simulating the Earth's atmosphere, ocean, land, land-ice, and sea-ice, plus one central coupler component. In this study, we mainly use the atmosphere component of CESM Version 1.0, i.e., the Community Atmosphere Model (CAM5) Version 5. This new version of the CAM 5.0 incorporates a number of enhancements to the physics package.

However, the representation of the microphysical processes of convective clouds in GCM is rather unrealistic. Since the vertical velocity and latent heating within deep convective cloud (DCC) systems are crudely diagnosed in the traditional convective parameterizations, the pathways of aerosols to interact with DCC and convective precipitation are excluded in the physics and dynamics of GCMs. Therefore, GCMs can only investigate the aerosol-indirect effects for stratiform/cirrus clouds, but are unsuitable for convective clouds associated with monsoon systems and storm track. There are possible solutions to tackle this dilemma. In this study, we introduce a hierarchical modeling approach with the combination of the models with different scales, i.e., to upscale aerosol forcings calculated from the cloud-resolving model results to the global simulations. A detailed discussion of this method will be presented in Chap. 5.

2.3 Multiscale Aerosol-Climate Modeling Framework

To overcome the deficiencies inherent in the traditional GCMs, the multiscale modeling framework (MMF) was introduced by Khairoutdinov and Randall (2001), which embeds a two-dimensional version of the cloud-resolving System for Atmospheric Modeling (SAM) at each grid column of a host NCAR CAM to resolve the subgrid variability in cloud dynamics and cloud microphysics and replace conventional parameterizations for moist convection and large-scale condensation. As an extension of the MMF, an updated aerosol-climate model (PNNL-MMF) was developed through upgrading the host GCM to CAM version 5, using an explicit-cloud parameterized-pollutant (ECPP) approach to link the cloud processing of aerosols on the large-scale grid with the cloud/precipitation statistics in the cloud-resolving model, and incorporating a two-moment microphysical scheme to replace the one-moment microphysics which was unable to simulate the interaction between aerosol particles and hydrometeors. More descriptions and evaluations of PNNL-MMF can be found in Wang et al. (2011a, b).

References

Abdul-Razzak H, Ghan SJ, Rivera-Carpio C (1998) A parameterization of aerosol activation—1. Single aerosol type. J Geophys Res 103(D6):6123–6131

Dudhia J (1989) Numerical study of convection observed during the winter monsoon experiment using a mesoscale two-dimensional model. J Atmos Sci 46(20):3077–3107

Fan J, Leung LR, Li Z, Morrison H, Chen H, Zhou Y, Qian Y, Wang Y (2012) Aerosol impacts on clouds and precipitation in eastern China: results from bin and bulk microphysics. J Geophys Res 117

Khairoutdinov M, Kogan Y (2000) A new cloud physics parameterization in a large-eddy simulation model of marine stratocumulus. Mon Weather Rev 128(1):229–243

Khairoutdinov MF, Randall DA (2001) A cloud resolving model as a cloud parameterization in the NCAR community climate system model: preliminary results. Geophys Res Letts 28 (18):3617–3620

Khain A, Ovtchinnikov M, Pinsky M, Pokrovsky A, Krugliak H (2000) Notes on the state-of-the-art numerical modeling of cloud microphysics. Atmos Res 55(3–4):159–224

Khain A, Pokrovsky A, Pinsky M, Seifert A, Phillips V (2004) Simulation of effects of atmospheric aerosols on deep turbulent convective clouds using a spectral microphysics mixed-phase cumulus cloud model. Part I: Model description and possible applications. J Atmos Sci 61(24):2963–2982

Khain A, Rosenfeld D, Pokrovsky A (2005) Aerosol impact on the dynamics and microphysics of deep convective clouds. Q J R Meteorol Soc 131(611):2639–2663

Khain AP, BenMoshe N, Pokrovsky A (2008) Factors determining the impact of aerosols on surface precipitation from clouds: an attempt at classification. J Atmos Sci 65(6):1721–1748

Khain AP, Leung LR, Lynn B, Ghan S (2009) Effects of aerosols on the dynamics and microphysics of squall lines simulated by spectral bin and bulk parameterization schemes. J Geophys Res 114(D22). doi:10.1029/2009JD011902

Kovetz A, Olund B (1969) Effect of coalescence and condensation on rain formation in a cloud of finite vertical extent. J Atmos Sci 26(5P2):1060–1065

Li G, Wang Y, Zhang R (2008) Implementation of a two-moment bulk microphysics scheme to the WRF model to investigate aerosol-cloud interaction. J Geophys Res 113(D15):D15211

Li G, Wang Y, Lee K-H, Diao Y, Zhang R (2009) Impacts of aerosols on the development and precipitation of a mesoscale squall line. J Geophys Res 114(D17):D17205

Liu YG, Daum PH (2004) Parameterization of the autoconversion process. Part I: Analytical formulation of the Kessler-type parameterizations. J Atmos Sci 61(13):1539–1548

Liu YG, Daum PH, McGraw R (2004) An analytical expression for predicting the critical radius in the autoconversion parameterization. Geophys Res Lett 31(6)

Lynn BH, Khain AP, Dudhia J, Rosenfeld D, Pokrovsky A, Seifert A (2005) Spectral (Bin) microphysics coupled with a mesoscale model (MM5). Part II: Simulation of a CaPE rain event with a squall line. Mon Weather Rev 133(1):59–71

Morrison H, Curry JA, Khvorostyanov VI (2005) A new double-moment microphysics parameterization for application in cloud and climate models. Part I: Description. J Atmos Sci 62(6):1665–1677

Morrison H, Grabowski WW (2007) Comparison of bulk and bin warm-rain microphysics models using a kinematic framework. J Atmos Sci 64(8):2839–2861

Pruppacher HR, Klett JD (1997) Microphysics of clouds and precipitation. Oxford Press, Oxford, p 914

Solomon A, Morrison H, Persson O, Shupe MD, Bao JW (2009) Investigation of microphysical parameterizations of snow and ice in arctic clouds during M-PACE through model-observation comparisons. Mon Weather Rev 137(9):3110–3128

Wang M et al (2011a) The multi-scale aerosol-climate model PNNL-MMF: model description and evaluation. Geosci Model Dev 4(1):137–168

Wang M, Ghan S, Ovchinnikov M, Liu X, Easter R, Kassianov E, Qian Y, Morrison H (2011b) Aerosol indirect effects in a multi-scale aerosol-climate model PNNL-MMF. Atmos Chem Phys 11(11):5431–5455

Chapter 3
Impacts of Urban Pollution on Thunderstorms

Abstract In this study, we have performed an analysis of lightning, precipitation, and visibility and numerical modeling to elucidate the relationship between air pollution and thunderstorms and to quantify the aerosol indirect effects on cloud development, precipitation, and lightning over the PRD area. Seven year measurements of precipitation, lightning flashes, and visibility from 2000 to 2006 have been analyzed in this area. To assess the effects of aerosols on cloud processes, precipitation, lightning activity, a WRF model with a two-moment bulk microphysical scheme has been employed to simulate a mesoscale convective system in this area. Sensitivity experiments have been performed to reflect aerosol conditions characteristic of both polluted and clean cases to further reveal the physical mechanism for the precipitation and lightning enhancement under the polluted aerosol condition. Note that the present modeling work only focuses on the indirect effect of aerosols, by which aerosols serve as CCN and hence affect precipitation and lightning activities of the thunderstorm event.

Keywords Aerosols · Thunderstorm · Heavy precipitation · Lightning

3.1 Long-Term Observations of Precipitation, Lightning Flashes, and Visibility

Measurements of precipitation, lightning flashes, and visibility from 2000 to 2006 are analyzed to investigate the relationship between rainfall, lightning frequency, and aerosols over the PRD area. The annual mean AOD from MODIS provides an overview of the aerosol distribution over the Guangdong Province. High-resolution (1 km) AOD at the 550 nm channel over the PRD area has been derived from MODIS by the Hong Kong University of Science and Technology (Li et al. 2005). Figure 3.1a depicts the annual mean AOD in 2005, showing that the Guangzhou megacity area has a higher aerosol loading with the AOD value larger than 0.6,

© Springer-Verlag Berlin Heidelberg 2015   17
Y. Wang, *Aerosol-Cloud Interactions from Urban, Regional, to Global Scales*,
Springer Theses, DOI 10.1007/978-3-662-47175-3_3

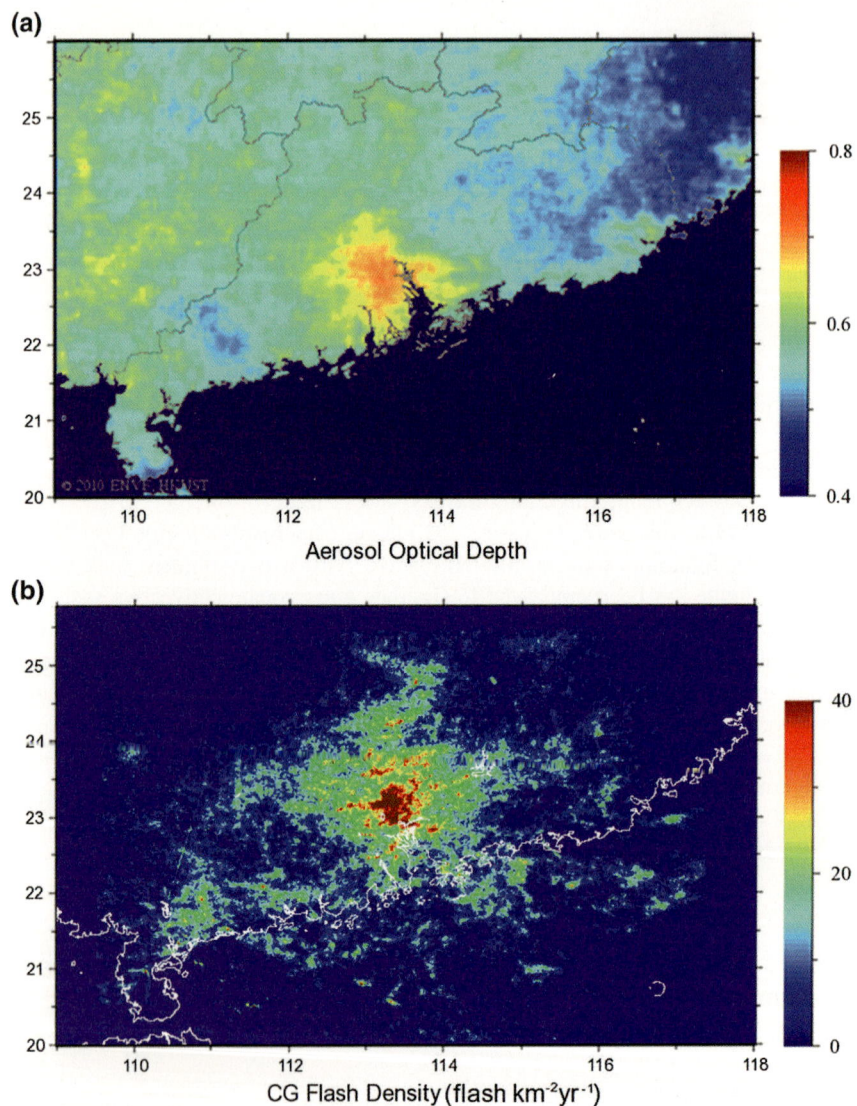

Fig. 3.1 AOD and flash density distribution over Southern China and flash density distribution over Southern China. **a** Annual mean AOD from the MODIS satellite in 2005. **b** Annual mean CG flash density distribution from the local lightning detection network in the Guangdong Province in 2005 (Reprinted from Wang et al. (2011) with permission of Copernicus Publications)

compared with the vicinity of the PRD region, consistent with the previous studies (Wu et al. 2005).

Lightning flashes are taken from a local lightning monitoring system, which has been established since 2000. This represents a province-wide lighting detection

(LD) network consisting of 16 lightning sensors to provide the coverage of lightning flashes for the entire Guangdong Province, with a focus on the Guangzhou megacity area. The LD network detects positive and negative cloud-to-ground lightning with a positioning error of less than 1 km and measures the electricity current associated with each flash. Figure 3.1b displays the annual cloud-ground (CG) flash density distribution in 2005 over the Guangdong Province. The lightning flashes exhibit a high flash density in the Guangzhou megacity area, with the highest flash density of 40 flashes km^{-1} $year^{-1}$. A comparison between the lightning flashes and AOD shows a similarity in the geographic distributions of the two quantities over the PRD area, with the largest values coinciding with the Guangzhou megacity area.

Atmospheric visibility is affected by many factors, such as absorbing gaseous pollutants and concentrations, distributions and chemical composition of aerosols. In a homogeneous atmosphere, the observed visibility (R) is correlated with the extinction coefficient (β) through the Koschmieder formula: $R = 3.91 \beta^{-1}$ (Seinfeld and Pandis 2006). Several previous studies have suggested that the PM amount is well correlated to visibility with a correlation coefficient of above 0.8 over the PRD area (Wu et al. 2005; Deng et al. 2008). Over the PRD area, visibility measurements are based on daily observations in four cities, including Guangzhou, Shenzhen, Zengcheng, and Huiyang. The daily visibility value is averaged from four measurements per day. We have excluded the days with precipitation in producing the statistics of the daily mean visibility. The number of low visibility days caused by light fog (relative humidity (RH) > 90 %), which is also excluded from the present work, is less than 3 days per month over the PRD area since 2000, as previously reported by Wu et al. (2007). It should be pointed out that the use of visibility as a proxy for the aerosol content has certain uncertainties. In particular, the aerosol optical properties may also be dependent on the RH, since hygroscopic aerosols will increase their size as RH increases (Zhang et al. 2008a). Nevertheless, RH measurements have been examined from 2000 to 2006 and the daily averaged RH (in exclusion of rainy and foggy days) typically ranged from 65 to 70 %, indicating that the interannual variation of visibility in the data cannot be explained by the variation of RH. Hence, the visibility results used in this study directly correlate with the aerosol loading condition in the atmosphere.

The correlation of heavy precipitation and lightning with visibility averaged over 4 months (March–June) from 2000 to 2006 is displayed in a scatter plot (Fig. 3.2). In the present study, we focus on heavy precipitation, which is defined as a daily rainfall amount greater than 25 mm. To exclude the seasonal factors contributing to the variations in precipitation and lightning, such as more frequent intrusions of tropical cyclones and larger intensity of solar heating induced convection in the summer, we focus on the period from March to June during the 7 years. Figure 3.2a illustrates that daily heavy rainfall is inversely correlated with visibility; the linear Pearson correlation coefficient is –0.739. Similarly, Fig. 3.2b exhibits a negative correlation between lightning flashes density and visibility, with a correlation coefficient of –0.506. Hence, the analysis of the 7 year measurements of precipitation, lightning flashes, and visibility from 2000 to 2006 in the PRD region

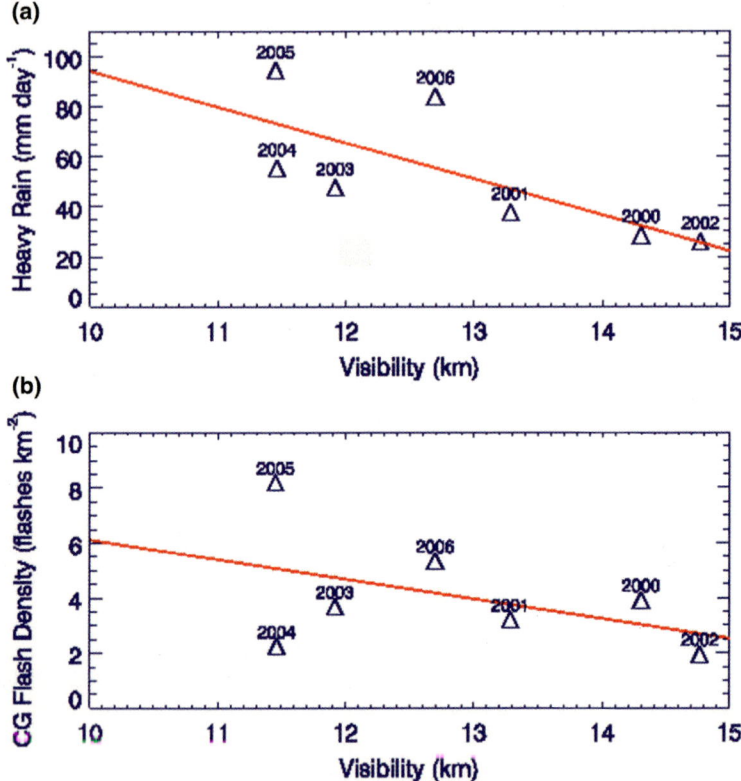

Fig. 3.2 Correlation between **a** daily visibility and the heavy rainfall rate, **b** daily visibility and lightning flash density over PRD from 2000 to 2006. Both daily visibility and heavy rainfall rate are averaged over the 4-month period (from March to June). The heavy rainfall rate is calculated from each rain gauge station with a daily rainfall greater than 25 mm. CG lightning flash density (flashes km^{-2}) is accumulated over 4 months each year. The *line* represents a linear regression through all data (Reprinted from Wang et al. (2011) with permission of Copernicus Publications)

indicates that the large lightning density and heavy rainfall amount in the PRD area are closely linked with atmospheric aerosol loading and local anthropogenic pollution contributes to the occurrences of extreme weather events, including lightning and heavy rainfall.

3.2 Design of Numerical Simulations

To further elucidate the effects of aerosols on cloud processes, precipitation, and lightning activity, simulations using the CR-WRF model with the two-moment bulk microphysical scheme are conducted. A mesoscale convective system over the PRD area associated with a cold frontal passage on March 28, 2009 is simulated in this study. The development of a frontal system is captured by infrared images of clouds in terms of the brightness temperature variance of the cloud top. At 0500 UTC on March 28, an embryonic convective cell appeared to the west of the Guangdong province. The convective system intensified and progressed eastwards, and at 0900 UTC a mesoscale convective system swept through the PRD urban area. The extended planetary boundary layer due to the surface roughness and heat island effect in the urban region and high aerosol loading likely enforced the vertical motion of the system, responsible for the occurrence of the low brightness temperature (<65 °C). From the atmospheric sounding profile over Guangzhou city (23.2° N, 113.3° E), the CAPE was estimated to be 1055 J Kg^{-1} at 0000 UTC, exhibiting an unstable thermodynamic condition for thunderstorm initialization along the frontal boundary.

The CR-WRF model is configured by a two-way interactive nested domain with two meteorology input files from 1800 UTC March 27 to 1800 UTC March 28, 2009. The outer domain has the size of 900 km × 600 km with a 3 km horizontal resolution and is centered at (23.5° N, 114° E). The nested domain provides a finer resolution of 1 km and covers a 300 km × 300 km area, centering at (22.8° N, 113.5° E), as illustrated in Fig. 3.3. The 1° × 1° NCEP Final Global Analyses data is used to establish the initial conditions and boundary conditions. To accurately represent city-scale surface physical processes over the urban groups in the PRD area, the Noah land surface model (LSM) coupled with the single-layer urban canopy model (Kusaka et al. 2010) is utilized in the CR-WRF simulation. The high-resolution land-use data from the Guangdong local Geographic Information System (GIS) database is adopted in the domain initialization, considering the large impact of the urban land use on the convection development (van den Heever and Cotton 2007). The Yonsei University

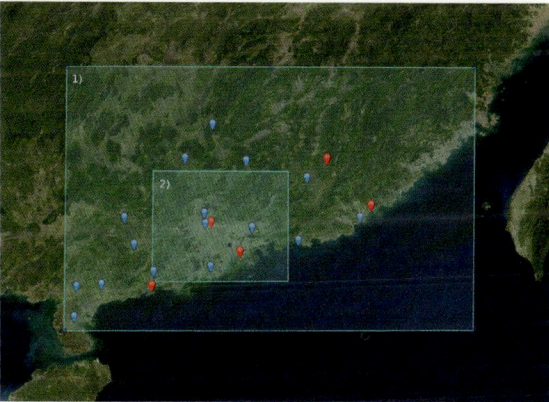

Fig. 3.3 Overview of domains in the model simulations. The *red symbols* represent the radar stations, and the *blue symbols* represent the lightning detection stations (Reprinted from Wang et al. (2011) with permission of Copernicus Publications)

(YSU) scheme is used to parameterize the boundary layer processes (Hong et al. 2006). No cumulus cloud parameterization is involved in the simulation.

To investigate the aerosol effect on thunderstorm and lightning, we conduct sensitive experiments with two aerosol scenarios: a polluted case (P-case) and a clean case (C-case), on the basis of atmospheric measurements conducted in this region (Zhang et al. 2008b). Both cases contain two aerosol types, i.e., ammonium sulfate and sea salt. The identical dynamics and thermodynamics settings in the sensitivity studies preclude the impacts on the simulations from other factors, such as large-scale dynamics, solar heating, and humidity sources. According to the Köhler equation used for aerosol activation in the model, the difference in the chemical composition of aerosols is equivalent to a corresponding change in the CCN effective radius (Khain et al. 2000). The polluted continental aerosols are assumed to mainly contain ammonium sulfate and the clean aerosols are assumed to be mainly sea salt (Seinfeld and Pandis 2006). Hence, ammonium sulfate and sea salt represent dominantly the fine mode and coarse mode in the aerosol spectrum, respectively. Background ammonium sulfate over the continent in the P-case is initiated as an exponentially decreasing profile with the maximal number concentration of $2200 \ cm^{-3}$ and a mass mixing ratio of $5.9 \ \mu g \ Kg^{-1}$ at the surface, while the initial aerosol profile of ammonium sulfate in the C-case is smaller, with the maximum value of $220 \ cm^{-3}$. The number concentration and mass mixing ratio of ammonium sulfate aerosol over the surface in the polluted case are constrained by field measurements from Liu et al. (2008a, b), respectively, in the field campaign PRIDE-PRD 2004. The exponential decreasing profile of aerosol vertical distribution is supported by aircraft measurement during PRIDE-PRD 2004 (Wang et al. 2008). The most recent analysis from high-resolution aerosol mass spectrometer (AMS) measurements (He et al. 2011) provides a similar mass concentration of ammonium sulfate to the value employed in this study. The mass and number concentration of ammonium sulfate over the sea are assumed to be half of the values over the continent. To mimic urban aerosol pollution, a production rate of ammonium sulfate is assumed to be $0.5 \ \mu g \ Kg^{-1} \ h^{-1}$ over the PRD metropolitan area (Zhang et al. 2008b). The background sea salt is initiated with a maximum number concentration of about $110 \ cm^{-3}$ and a mass mixing ratio of $0.36 \ \mu g \ Kg^{-1}$ in both P- and C-cases. A sea salt emission scheme considering wind velocity and relative humility over the sea surface is included in the model (Li et al. 2008). Activation process represents the only sink for aerosols considered in our current microphysical scheme.

3.3 Model Evaluation and Sensitivities for Aerosol Levels

CR-WRF simulations are evaluated through the comparison between the measured and modeled distributions of the maximum radar reflectivity at two time periods in Fig. 3.4. To derive the radar reflectivity, both the simulated mass and number concentration of raindrops, snow and graupel are considered to calculate the six-moment

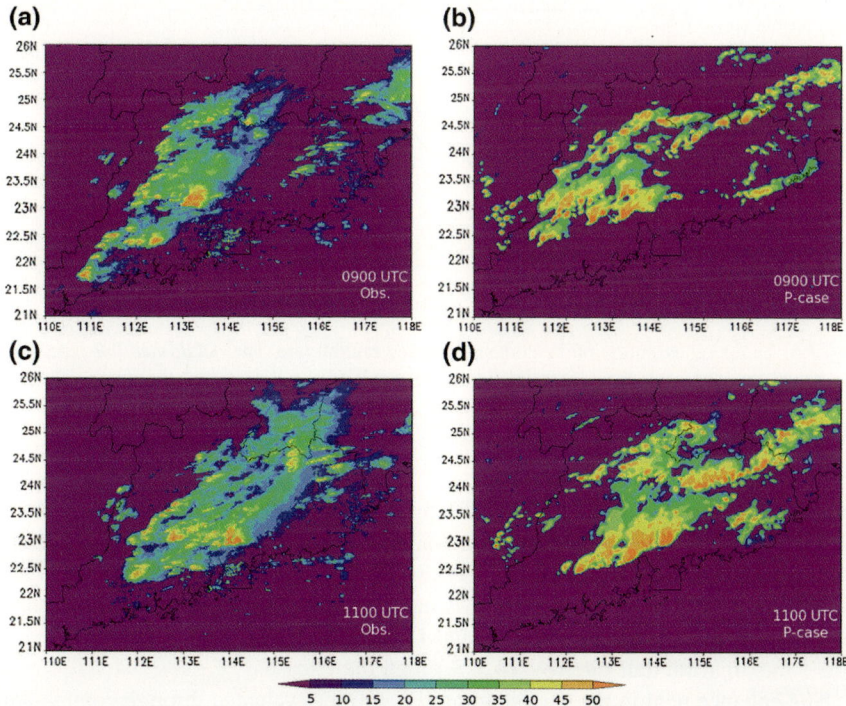

Fig. 3.4 Comparison of radar reflectivity between observation and CR-WRF simulation for the P-case. **a** Observation at 0900 UTC. **b** Simulation at 0900 UTC. **c** Observation at 1100 UTC. **d** Simulation at 1100 UTC (Reprinted from Wang et al. (2011) with permission of Copernicus Publications)

of hydrometeors, which is defined as the reflectivity factor. Comparison of southwest orientated convective boundary along the frontal system. Most of the radar reflectivity is made between model simulations and observations from five S-band (10 cm) Doppler radars (the locations of the radars are marked in Fig. 3.3). Comparison between the observation and simulation exhibits a general agreement in the distributions and developments of the maximum radar reflectivity, showing radar reflectivities along the frontal boundary are reproduced by the simulation. In particular, the simulation is consistent with the observation on the location of the reflectivity value greater than 30 dBz, corresponding to the location of active convection development of the thunderstorm with heavy precipitation in the north of the PRD area between latitudes 23° N and 23.5° N.

Precipitation measurements from 2000 gauge stations in the Guangdong province are compared with the WRF simulation. Figure 3.5 displays hourly rainfall rate horizontally averaged over the PRD area from the gauge measurements and from simulations for the C-case and P-case. The simulated precipitation in the P- and C-cases occurs about 1 h earlier than the gauge measurement, but the precipitation

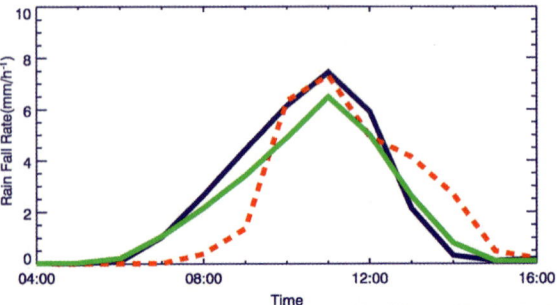

Fig. 3.5 Temporal evolution of the domain-averaged rainfall rate. The *red dashed line* represents the gauge measurement, the *blue solid line* corresponds to the P-case, and the *green solid line* corresponds to the C-case (Reprinted from Wang et al. (2011) with permission of Copernicus Publications)

rate reaches the peak value at 0900 UTC in both simulations and the measurement. The maximal rainfall rate in the P-case is consistent with the gauge measurement, with a value of 7.6 mm h^{-1} at 0900 UTC, while the peak rainfall rate of 6.5 mm h^{-1} is noticeably smaller in the C-case. It is interesting to note that, although the onset of precipitation between the P-case and C-case is similar, the P-case produces a larger maximal rainfall rate (by 14.5 %) than the C-case.

The enhancement in the surface rainfall is further evaluated through comparison of the spatial distribution of precipitation. Figure 3.6 depicts the accumulated precipitation from UTC 0900 to UTC 1500 over the PRD area. Both the P-case and C-case reproduce the precipitation belt to the north part of the PRD area, while the coverage of the simulated precipitation belts extend slightly to the south compared with the gauge measurement. A comparison between the P-case (Fig. 3.6b) and C-case (Fig. 3.6c) reveals that precipitation increases in most parts of the region under the high aerosol loading. Note that the enhancement in precipitation is non-uniform over the entire domain. For instance, in the C-case the region with

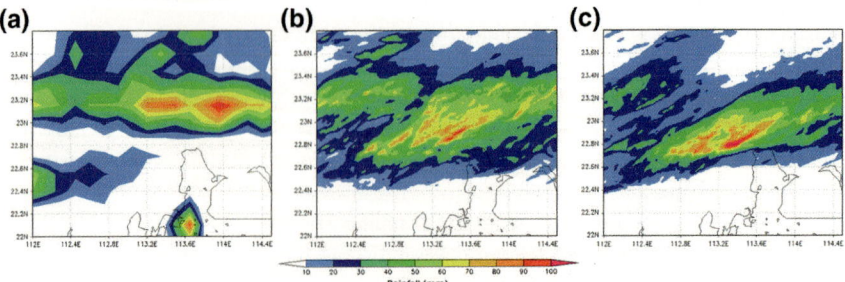

Fig. 3.6 Comparison of accumulated precipitation from 0900 to 1500 UTC between observation and simulations. **a** Gauge measurements. **b** P-case. **c** C-case (Reprinted from Wang et al. (2011) with permission of Copernicus Publications)

accumulated precipitation greater than 100 mm inside the main rain belt is slightly larger than that in the P-case. In terms of the total accumulated rainfall integrated through the duration of the thunderstorm, the P-case and C-case produce 30.5 and 26.4 mm total precipitation, respectively, over the out domain. Both values are comparable with the corresponding gauge measurement of 27.9 mm, but the P-case yields a value of 16 % greater than that in the C-case. Hence, the model simulations reveal that elevated aerosol concentrations increase precipitation associated with the mesoscale convective system over the PRD area.

We examine the detailed precipitation response to aerosols by considering several distinct precipitation categories. The precipitation is categorized into four levels by the daily precipitation amount, i.e., light (<10 mm d^{-1}), moderate (10–25 mm d^{-1}), heavy (25–50 mm d^{-1}), and extreme heavy (>50 mm d^{-1}) rain. Figure 3.7 shows the percentage of the grid areas under a certain precipitation category over the entire model domain (a) and the percentage of the precipitation amount under a certain category over the total precipitation amount (b). It is evident that light rain is distributed over lesser geographic grid areas and the percentage of light rain in the total precipitation amount also decreases in the P-case, revealing a reduction in light precipitation because of elevated aerosol loading. This result is in accordance to a previous study by Qian et al. (2009), which suggested that heavy pollution in China suppresses light rain on the basis of observation and numerical modeling. Our results show that the heavy rain and extremely heavy rain are both enhanced in the total precipitation amount and in the geographic distribution for the P-case. Hence, the model simulations indicate that while elevated aerosol loading suppresses light and moderate precipitation, it enhances heavy precipitation.

3.4 Lightning Flashes and Lightning Potential Index

Lightning occurrence and frequency are often considered as an important indicator of convective intensity, since charge separation and electrification of thunderstorms require the coexistence and interaction of both supercooled liquid and ice crystals (Williams et al. 1991). Mesoscale convective systems produce ground flashes by an average rate of 42 flashes min^{-1}, which consists of about one-fourth of the annual lightning strikes globally (Goodman and MacGorman 1986). Considering the complicated microphysical processes involved in the electrical field buildup and lightning discharge, it is difficult to exactly represent the lightning occurrence in atmospheric models. In the present work, we evaluate the lightning activity associated with the mesoscale convective system on the basis of a lightning potential index (LPI), which has been developed by Yair et al. (2010). The LPI parameterizes the potential of charge generation and separation in convective thunderstorm on the basis of the noninductive graupel–ice mechanism, since there is a growing consensus that the occurrence of electrical charge transfer when graupel particles in the region of intense updraft collide with small ice crystal represents the dominant electrification mechanism in thunderstorms (Rakov and Uman 2003). Under the assumption

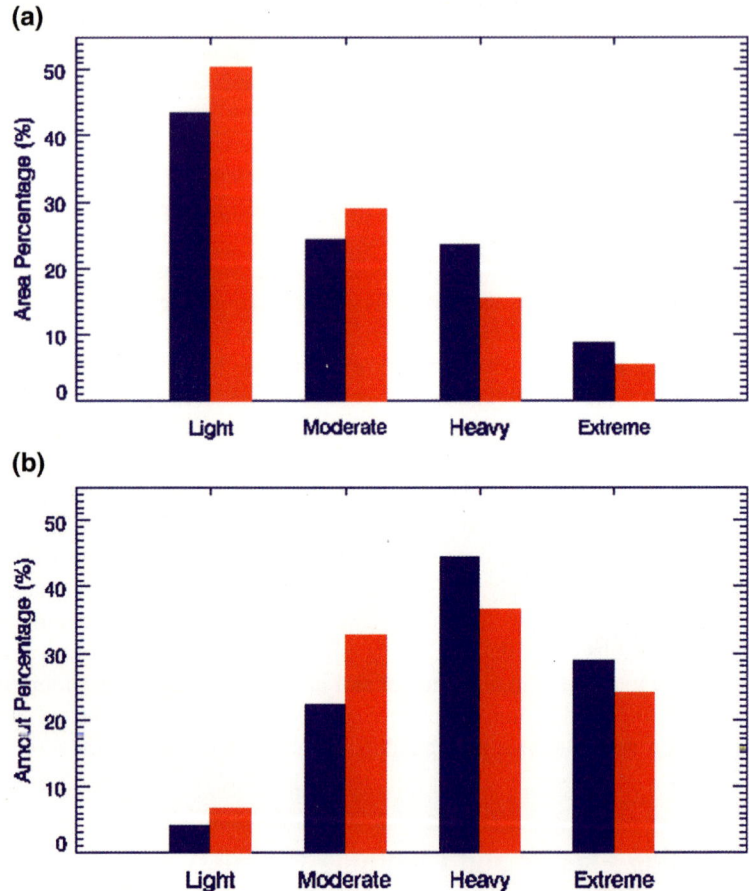

Fig. 3.7 Probability distribution functions of the four different rainfall categories for the P- and C-cases. **a** Percentage of grid areas under a certain precipitation category over the entire model domain. **b** Percentage of the precipitation amount under a certain category over the total precipitation amount. *Dark blue* corresponds to the P-case, and *red* corresponds to the C-case (Reprinted from Wang et al. (2011) with permission of Copernicus Publications)

that the effective charge separation zone is between freezing level and −20 °C isotherms, determination of LPI mainly involves the simulated vertical wind component, temperature field, and the mass mixing ratios of hydrometeors explicitly predicted from the microphysical scheme, i.e., LPI $= 1/V \int \int \int \varepsilon w^2 \mathrm{d}x\mathrm{d}y\mathrm{d}z$, where ε is decided by the partition of water content between liquid phase and ice phase. This method has been proved to be more accurate than other thermodynamic indices for lightning, such as the K-index and lifted index (Yair et al. 2010).

    The temporal evolution of measured flashes from the local lightning detection network (the locations of the lightning sensors are marked in Fig. 3.3) is shown in Fig. 3.8a, including both positive and negative flashes. Figure 3.8a indicates that

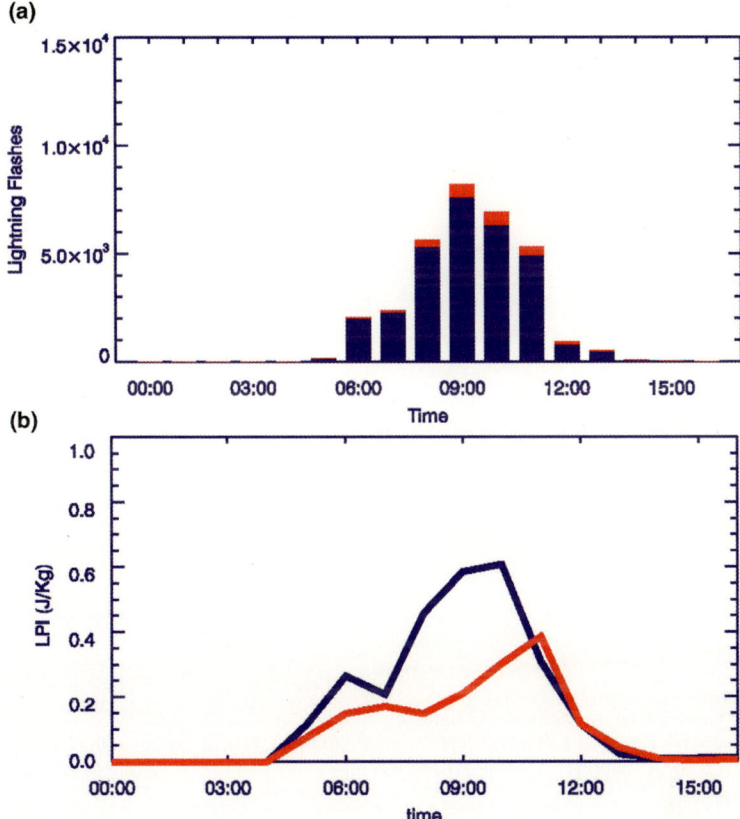

Fig. 3.8 Temporal evolution of **a** observed total CG lightning flashes and **b** calculated domain-averaged LPI. In **a** *red* denotes positive flashes and *dark blue* denotes negative flashes. In **b** *red* denotes the C-case and *dark blue* denotes the P-case (Reprinted from Wang et al. (2011) with permission of Copernicus Publications)

positive flashes dominate over the negative flashes and the occurrence of lightning flashes reaches the peak at 0900 UTC. The time dependence LPI averaged over PRD calculated under the P- and C-cases is shown in Fig. 3.8b. Compared to the C-case, the evolution of LPI in the P-case exhibits a better consistency with the measured lightning flashes over PRD. The value of calculated LPI for the P-case and measured flashes are positive at 0400 UTC, increase significantly after 0700 UTC, and reach the peak around 0900 UTC. In contrast, the predicted LPI for the C-case is noticeably delayed, compared to the P-case and measured lightning flashes. Over the entire thunderstorm duration LPI in the P-case is about 53 % higher than that in the C-case.

Figure 3.9 displays measured 1 h lightning flashes from 0800 to 0900 UTC, along with the calculated LPI at 0800 UTC in the two cases. The observation shows the highest frequency of lightning occurs to the west of Guangzhou City at 0800 UTC and progressed slightly southeastward at 0900 UTC. The P-case reproduces

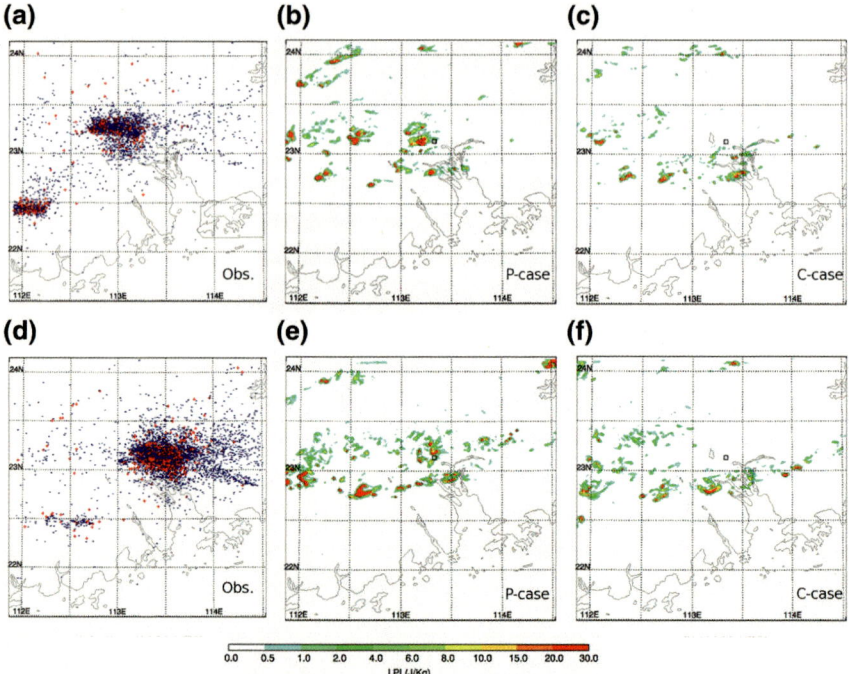

Fig. 3.9 Comparison of observed CG lightning distribution with the simulated LPI in the P-case. **a** Strikes (0800 UTC), **b** LPI (0800 UTC), **c** LPI (0800 UTC), **d** Strikes (0900 UTC), **e** LPI (0900 UTC), **f** LPI (0900 UTC) (Reprinted from Wang et al. (2011) with permission of Copernicus Publications)

these measured characteristics, showing that the area of the LPI values larger than 1.0 locates to the west of Guangzhou and migrates to the Guangzhou metropolitan and PRD area at 0900 UTC. Such a feature, however, is absent from the calculated LPI for the C-case. It is also evident in Fig. 3.9 that there are discrepancies in the geographic distributions between the measured lightning activities and predicted LPI values to the northwest of PRD, which may be jointly explained by the limitation in the lightning detection system (which detects the cloud-to-ground flashes only), uncertainties in the model calculations, and the imperfect nature of the lightning potential parameterization.

3.5 Microphysical Properties and Convections

We analyze the modeled microphysical properties under the different aerosol conditions to gain insights into the effects of aerosols on cloud development, precipitation, and lightning activity. Table 3.1 summarizes the domain-averaged

Table 3.1 Domain-averaged properties of hydrometeors in the CR-WRF simulations

	Number concentration (10^3 m^{-3})		Effective radius (μm)	
	C-case	P-case	C-case	P-case
Cloud droplets	1354	7663	7.8	6.0
Rain drops	2.28	1.17	372.4	574.3
Ice crystals	1.6	3.3	226.0	227.4
Snow	0.6	0.8	294.8	292.9
Graupel	2.0	1.2	445.4	660.6

microphysical properties for the different hydrometeors. For the P-case, the number concentration of cloud droplets is much larger, but the effective radius of cloud droplets is smaller than those for the C-case. These differences in the properties of cloud droplets are a direct reflection of the aerosol condition, since elevated aerosol loading leads to more numerous cloud droplets, but a smaller effective radius in the P-case. In addition, the number concentration of rain drops is slightly larger in the C-case, but the P-case corresponds to a much larger size for rain drops. Since a smaller effective radius of cloud droplets suppresses the collection/coalescence process to form raindrops, this likely explains the lower concentration of rain drops in the P-case. The larger size of rain drops in the P-case is indicative of the contributions from melting graupel. It is also clear from Table 3.1 that the concentrations of ice crystals and snow are higher in the P-case, suggesting a hindered warm rain process, but an enhanced mixed process because of elevated aerosol loading.

The temporal variations of the vertical profiles of the four hydrometeors (i.e., cloud water, rain water, ice, and graupel) are displayed in Fig. 3.10. For each quantity, the mass mixing ratio of hydrometeors is integrated horizontally at a given altitude. Figure 3.10a, b show shat that the amount of liquid cloud droplets is increasingly produced, when convection is initialized at 0400 UTC in both P- and C-cases. This occurs since supersaturation occurs due to adiabatic cooling in the ascending air mass. The P-case yields considerably more liquid cloud water in the lower troposphere than the C-case; between 0600 and 1400 UTC the amount of liquid cloud water in the P-case sustains higher than that in the C-case. For the P-case, more aerosols are activated into cloud droplets and more water vapor condenses onto cloud droplets in the P-case partially explaining the enhanced cloud water amount. Furthermore, a smaller effective radius of cloud droplets and suppressed collection/coalescence processes leads to a less efficient sedimentation for smaller cloud droplets and a prolonged condensation process in the atmosphere. Clearly, the more abundant cloud water in the P-case resulting in a larger release of latent heat from droplet condensation, invigorating the convective development, is to be discussed later. The enhanced cloud water above the freezing level (around 4 km) in the P-case is explained because of enhanced convection and smaller cloud droplets, which are easier to be lifted up. More supercooled water and enhanced convection promote the mixed-phase processes. Consequently, ice and graupel

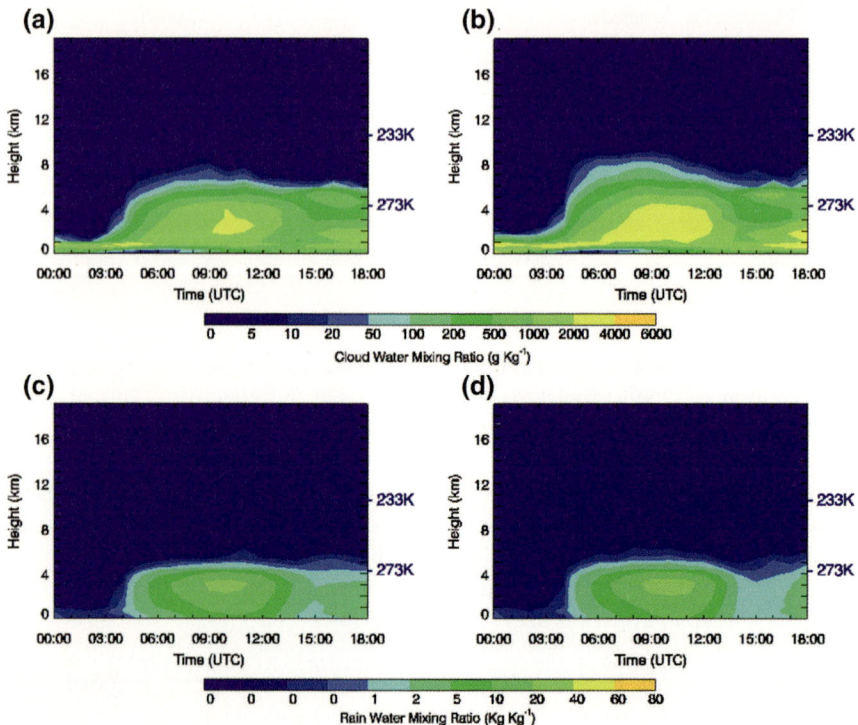

Fig. 3.10 Temporal evolution of the horizontally domain-summated mass mixing ratio of **a** cloud water in the C-case, **b** cloud water in P-case, **c** rain water in the C-case, **d** rain water in the P-case, **e** ice in the C-case, **f** ice in the P-case, **g** graupel in the C-case, and **h** graupel in the P-case (Reprinted from Wang et al. (2011) with permission of Copernicus Publications)

mass mixing ratios in the P-case are significantly increased compared to those in the C-case, as shown in Fig. 3.10e, f. Figure 3.10e shows that the region with ice crystals extends vertically to a higher altitude, indicating an elevated cloud top height. The maximum ice content is located above the –40 °C level, implying that homogeneous freezing of supercooled droplets represents a dominant mechanism for ice initiation.

It is interesting to note that there is relatively little difference in the rainwater amount between the C-case (Fig. 3.10c) and P-case (Fig. 3.10d). Such a behavior is rather surprising, since it is anticipated that there will be less rainwater due to suppressed collection/coalescence processes in the P-case. However, the less efficient conversion from cloud droplets to rain drops may be compensated by melting graupel generated from the enhanced ice phase process. As discussed above, the contribution of melting graupel to rainwater is supported by the predicted larger size of rain drops (Table 3.1) in the P-case. In addition, a larger size of hydrometeors implies a better chance to survive evaporation during sedimentation below the

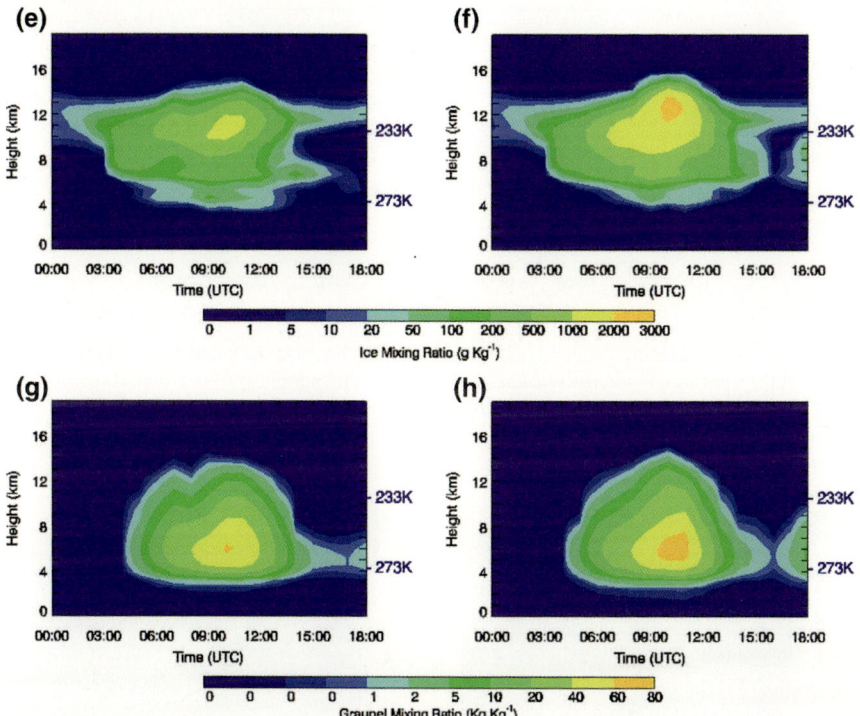

Fig. 3.10 (continued)

cloud bottom, explaining the intense rainfall period between 0900 and 1100 UTC at the ground level (0 km in the figure). Also, there is an insignificant difference in the rainfall onset, despite the suppressed warm rain process in the P-case.

Strong updraft (>10 m s^{-1}), which is characteristic of thunderstorms in the developing and mature stages, represents a dominant factor in the formation of heavy precipitation and lightning (Rakov and Uman 2003; Williams et al. 1991). We evaluate the vertical convection strength under different aerosol conditions to investigate the aerosols effect on cloud dynamics. Figure 3.11 shows the time series of the column maximum updraft and downdraft. The maximum upward and downward velocities are calculated at each column and averaged over the entire domain. It is evident from Fig. 3.11 that the variations in the updraft and downdraft are always in phase and both velocities reach the peak at 0900 UTC. The updraft and downdraft velocities are larger in the P-case.

The release of latent heat and its vertical distribution is of critical importance in the feedback of microphysics on the cloud dynamics (Seifert and Beheng 2005). Figure 3.12 shows the vertical profile of latent heat release. In the developing and mature stages of the mesoscale convective system, the latent heat release between 2 and 8 km is noticeably stronger in the P-case, because of a more efficient and

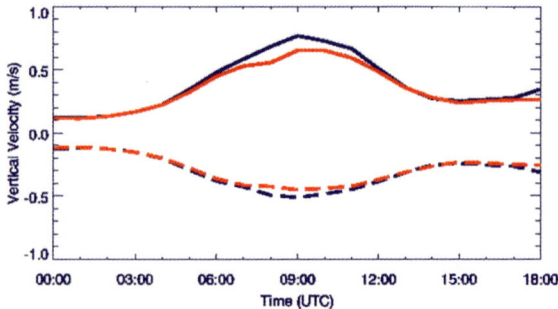

Fig. 3.11 Domain averaged column maximum vertical velocity in the simulations. The *solid lines* represent the updraft and the *dashed lines* represent the downdraft. The *dark blue lines* represent the P-case and the *red lines* represent the C-case (Reprinted from Wang et al. (2011) with permission of Copernicus Publications)

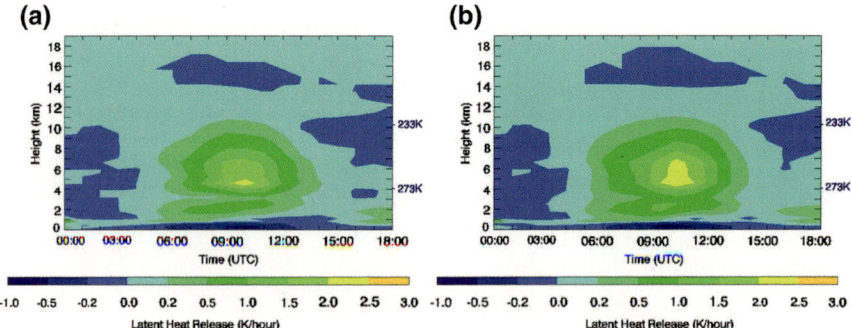

Fig. 3.12 Temporal evolution of latent heat profiles: **a** C-case and **b** P-case (Reprinted from Wang et al. (2011) with permission of Copernicus Publications)

prolonged condensation process, as discussed before. However, the cooling induced by evaporation and melting of hydrometeors below 2 km is not appreciably different between the two cases, indicating that the cold pool produced by evaporative cooling is less affected by aerosol leading. This vertical distribution of latent heat release contributes to destabilization of the atmosphere and further enhancement of the convective strength.

3.6 Summary

In this work, we present an analysis of long-term measurements of precipitation, visibility, and lightning in the Pearl River Delta (PRD) region, China, with a focus on the Guangzhou megacity area. Measurements of precipitation, lightning flashes, and visibility from 2000 to 2006 in this region are analyzed to assess the impacts of aerosols on cloud and precipitation processes. The statistical analysis shows that both heavy rainfall and lightning flashes over PRD are inversely correlated with visibility, with the correlation coefficients of −0.739 and −0.506, respectively. The results suggest that a large lightning flash density and a heavy rainfall amount in the PRD area may be correlated with atmospheric aerosol loading and local anthropogenic pollution contributes to the occurrences of extreme weather events, including lightning and heavy rainfall.

To further elucidate the effects of aerosols on cloud processes, precipitation, and lightning activity, the CR-WRF model with a two-moment bulk microphysical scheme is employed to simulate a mesoscale convective system occurring on March 28, 2009 in the Guangzhou megacity area. Comparison between observations and model simulations exhibits a general agreement in the distributions and developments of the maximum radar reflectivity. Most of the measured radar reflectivity along the frontal boundary is reproduced by the simulation, in particular, the location of the most active convection development of the thunderstorm. The modeled maximal rainfall rate and spatial distribution of precipitation are also in agreement with measurements from rain gauges over this region.

Model sensitivity experiments reveal that elevated aerosol concentrations increase precipitation associated with the mesoscale convective system over the PRD area. The simulations suggest that the total precipitation is enhanced by about 16 % under the polluted aerosol condition. The results also reveal that elevated aerosol loading suppresses light and moderate precipitation (less than 25 mm per day), but enhances heavy precipitation (greater than 25 mm per day). The LPI is considered to evaluate the lightning activity associated with the mesoscale convective system. The calculated LPI exhibits a temporal and spatial consistence with lightning flashes recorded by the local LD network. The simulations suggest that LPI is enhanced by about 50 % under the polluted aerosol condition.

An analysis of the microphysical properties of hydrometeors indicates that elevated aerosol loading yields more numerous cloud droplets, but a smaller effective radius, leading to a suppressed collection/coalescence process, but an enhanced mixed process. The model predicts considerably more cloud water in the lower troposphere for the polluted case, but indicates relatively little difference in the rainwater under different aerosol conditions. During the developing and mature stages of the mesoscale convective system, the latent heat release in the middle troposphere is enhanced under the polluted aerosol condition, because of a more efficient and prolonged condensation process, but the cold pool relevant to evaporative cooling is largely unaffected by the aerosol loading. Such a distribution of latent heat release may destabilize the atmosphere and enhance convection.

References

Deng X, Tie X, Wu D, Zhou X, Bi X, Tan H, Li F, Jiang C (2008) Long-term trend of visibility and its characterizations in the Pearl River Delta (PRD) region, China. Atmos Environ 42 (7):1424–1435

Goodman SJ, MacGorman DR (1986) Cloud-to-ground lightning activity in mesoscale convective complexes. Mon Weather Rev 114:2320–2328

He LY, Huang XF, Xue L, Hu M, Lin Y, Zheng J, Zhang RY, Zhang YH (2011) Submicron aerosol analysis and organic source apportionment in an urban atmosphere in Pearl River Delta of China using high-resolution aerosol mass spectrometry. J Geophys Res 116:D12304

Hong S-Y, Noh Y, Dudhia J (2006) A new vertical diffusion package with an explicit treatment of entrainment processes. Mon Weather Rev 134:2318–2341

Khain A, Ovtchinnikov M, Pinsky M, Pokrovsky A, Krugliak H (2000) Notes on the state-of-the-art numerical modeling of cloud microphysics. Atmos Res 55(3–4):159–224

Kusaka H, Kondo H, Kikegawa Y, Kimura F (2010) A simple single-layer urban canopy model for atmospheric models: comparison with multi-layer and slab models. Bound Layer Meteorol 101:329–358

Li C, Lau AK-H, Mao J, Chu DA (2005) Retrieval, validation, and application of the 1-km aerosol optical depth from MODIS measurements over Hong Kong. IEEE Trans Geosci Remote Sens 43(11):2650–2658

Li G, Wang Y, Zhang R (2008) Implementation of a two-moment bulk microphysics scheme to the WRF model to investigate aerosol-cloud interaction. J Geophys Res 113(D15):D15211

Liu S, Hu M, Slanina S, He LY, Niu YW, Bruegemann E, Gnauk T, Herrmann H (2008a) Size distribution and source analysis of ionic compositions of aerosols in polluted periods at Xinken in Pearl River Delta (PRD) of China. Atmos Environ 42(25):6284–6295

Liu S, Hu M, Wu ZJ, Wehner B, Wiedensohler A, Cheng YF (2008b) Aerosol number size distribution and new particle formation at a rural/coastal site in Pearl River Delta (PRD) of China. Atmos Environ 42(25):6275–6283

Qian Y, Gong D, Fan J, Leung LR, Bennartz R, Chen D, Wang W (2009) Heavy pollution suppresses light rain in China: observations and modeling. J Geophys Res 114:D00K02

Rakov VA, Uman MA (2003) Lightning: physics and effects. Cambridge University Press, Cambridge

Seifert A, Beheng KD (2005) A two-moment cloud microphysics parameterization for mixed-phase clouds. Part 2: Maritime vs. continental deep convective storms. Meteorology and Atmospheric Physics 92:67–82. doi:10.1007/s00703-005-0113-3

Seinfeld JH, Pandis SN (2006) Atmospheric chemistry and physics: from air pollution to climate change. Wiley, New York

van den Heever SC, Cotton WR (2007) Urban Aerosol Impacts on Downwind Convective Storms. Journal of Applied Meteorology and Climatology 46:828–850. doi:10.1175/jam2492.1

Wang W, Ren LH, Zhang YH, Chen JH, Liu HJ, Bao LF, Fan SJ, Tang DG (2008) Aircraft measurements of gaseous pollutants and particulate matter over Pearl River Delta in China. Atmos Environ 42(25):6187–6202

Wang Y, Wan Q, Meng W, Liao F, Tan H, Zhang R (2011) Long-term impacts of aerosols on precipitation and lightning over the Pearl River Delta megacity area in China. Atmos Chem Phys 11(23):12421–12436

Williams ER, Zhang R, Rydock J (1991) Mixed phase microphysics and cloud electrification. J Atmos Sci 48:2195–2203

Wu D, Tie X, Li C, Ying Z, Kai-Hon Lau A, Huang J, Deng X, Bi X (2005) An extremely low visibility event over the Guangzhou region: a case study. Atmos Environ 39(35):6568–6577

Wu D, Bi X, Deng X, Li F, Tan H, Liao G, Huang J (2007) Effect of atmospheric haze on the deterioration of visibility over the Pearl River Delta. Acta Meteorol Sinica 21(2)

Yair Y, Lynn B, Price C, Kotroni V, Lagouvardos K, Morin E, Mugnai A, Llasat MdC (2010) Predicting the potential for lightning activity in Mediterranean storms based on the Weather Research and Forecasting (WRF) model dynamic and microphysical fields. J Geophys Res 115 (D4):D04205

Zhang RY, Khalizov AF, Pagels J, Zhang D, Xue HX, McMurry PH (2008a) Variability in morphology, hygroscopicity, and optical properties of soot aerosols during atmospheric processing. Proc Natl Acad Sci USA 105(30):10291–10296

Zhang YH, Hu M, Zhong LJ, Wiedensohler A, Liu SC, Andreae MO, Wang W, Fan SJ (2008b) Regional integrated experiments on air quality over Pearl River Delta 2004 (PRIDE-PRD2004): overview. Atmos Environ 42(25):6157–6173

Chapter 4
Aerosol Effects on the Stratocumulus and Evaluations of Microphysics

Abstract This study aims at assessing the aerosol effects on the maritime strato-cumulus (Sc) clouds and improving the cloud microphysical schemes for more realistic simulations of aerosol and cloud properties in regional and climate models. The double-moment Morrison bulk microphysical scheme presently implemented in the WRF model (Morrison et al. 2005) has been modified by replacing the prescribed aerosols in the original bulk scheme (Bulk-OR) with a prognostic double-moment aerosol representation (Bulk-2M). Furthermore, the impacts of the parameterizations of droplet diffusional growth and autoconversion and the selection of the embryonic raindrop radius on the performance of the bulk microphysical scheme have also been evaluated by the field observations and spectral bin microphysics (SBM) simulations. A maritime Sc system has been investigated over the Southeast Pacific Ocean from an international program, VAMOS Ocean-Cloud-Atmosphere-Land Study (VOCALS), the major goal of which is to advance the scientific understanding of land-ocean-atmosphere coupled system over the Southeast Pacific region. The warm Sc system is selected because the microphysical and dynamical processes are relatively simple to identify the differences between bulk and SBM microphysics and to evaluate the causes and effects of different treatments in the microphysics.

Keywords Aerosol indirect effect · Cloud microphysics · Stratocumulus clouds

4.1 Experiment Design

The development of Sc and the formation of drizzle precipitation that occured on October 28, 2008 over the Southeast Pacific region from the VOCALS-Ex field campaign were simulated with WRF V3.1.1. Three nested domains are used in the model with the horizontal resolutions of 12, 3, and 1 km, as shown in Fig. 4.1a. The finest-grid domain covers the flight track of the NSF/NCAR C130 Research Aircraft on that day. Sixty-five vertical levels are used with a resolution of about 30 m within

© Springer-Verlag Berlin Heidelberg 2015   37
Y. Wang, *Aerosol-Cloud Interactions from Urban, Regional, to Global Scales*,
Springer Theses, DOI 10.1007/978-3-662-47175-3_4

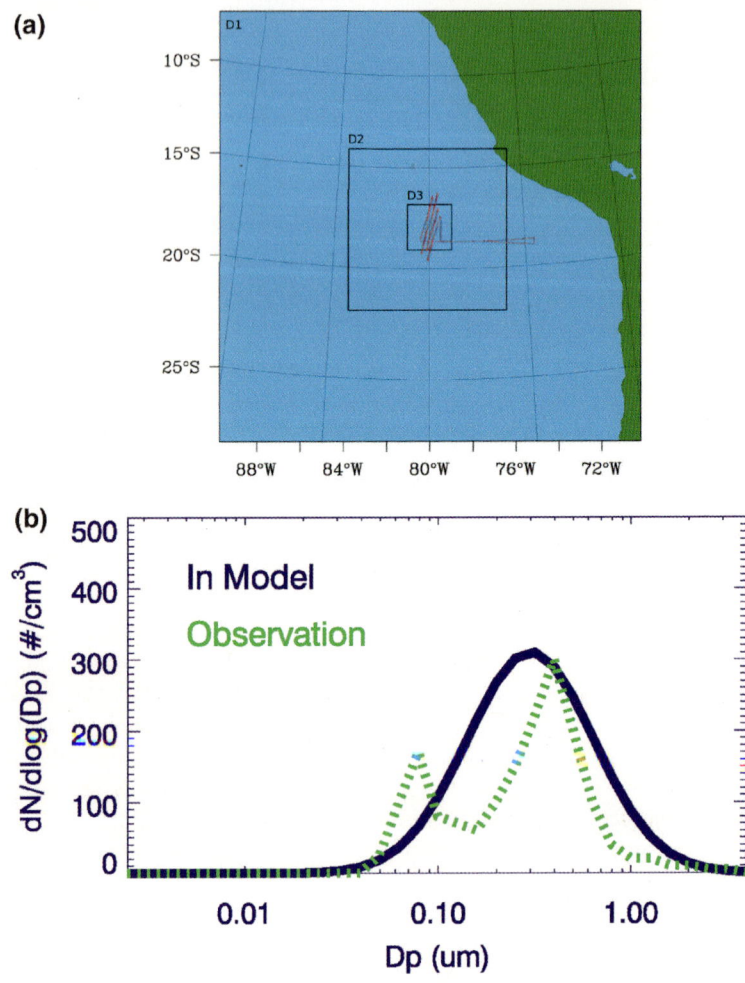

Fig. 4.1 a Domain overview for the simulation case. *Red line* denotes the flight track of C130 research aircraft on Oct 28, 2008. **b** Aerosol size distribution from field measurement and used in the model initialization (Reprinted from Wang et al. (2013) with permission of John Wiley and Sons)

the boundary layer. Such a high-vertical resolution is needed for realistic simulation of the moisture and supersaturation profile of stratocumulus clouds.

The Goddard shortwave radiation scheme and the RRTM longwave radiation scheme are employed in the simulations. Since the horizontal resolution of 1 km is still relatively coarse to resolve turbulent updrafts and eddies, subgrid-scale parameterizations play an important role in the Sc simulations (Yang et al. 2011). The Yonsei University (YSU) scheme is adopted to parameterize the planetary boundary layer processes (Hong et al. 2006). The heat eddy diffusion coefficient of the YSU scheme is adopted in the bulk microphysical scheme for the droplet activation calculation. The 'one-way nest down' technique is employed for the three domains, so that different microphysical schemes can be used on the domains with the same meteorological and aerosol conditions. Specifically, the simulations of the outer two domains are carried out first by using the Bulk-OR scheme to produce the initial and lateral boundary meteorological conditions for the finest-grid domain. Subsequently sensitive simulations with different configurations of microphysical schemes are conducted over the finest-grid domain.

The surface aerosol size distribution over the southeast Pacific region during the VOCALS-Ex field campaign was measured by a Differential Mobility Particle Sizer (DMPS) in combination with an Aerodynamic Particle Sizer (APS) aboard the NOAA RV Brown Research Ship. The ranges of particle diameter detected by DMPS and APS are 0.020–0.8 μm and 0.96–10 μm, respectively. On Oct 28, the ship routes were located in the south region within Domain 2. The measured size distribution of aerosols is shown in Fig. 4.1b. It is clear that aerosols in the Aitken and accumulation modes dominate the spectra and very few aerosol particles exist in the coarse mode. Hence, a single log-normal curve is applied to fit the observational data, yielding a total aerosol number concentration of 588 particles/cm^3 to be consistent with the observation. The aerosol mass mixing ratio is 2.4×10^{-11} g/cm^3. Ammonium sulfate is assumed to be the principal chemical composition of aerosols, although other constituents, particularly organics can also be present in ambient aerosols (Zhao et al. 2006; Wang et al. 2010). Aerosols in both the SBM and bulk schemes are initialized by the same spectra. Compared to typical maritime conditions over the southeast Pacific region, the above aerosol conditions (also referred to as the control case) correspond to elevated aerosol loading, which is likely under the influence of continental flows (Yang et al. 2011).

4.2 Effects of Aerosol Representation on Sc Simulations

4.2.1 Simulated Aerosol Evolution

The simulated aerosol evolutions from three microphysical schemes (Bulk-OR, Bulk-2M and SBM) are presented in Fig. 4.2. In the first few hours of the simulations, the large difference of the aerosol concentrations between Bulk-OR and Bulk-2M occurs, which is attributed to the efficient nucleation scavenging in the Bulk-2M and SBM under the favorable meteorological condition to form clouds within the maritime boundary layer. Since the total aerosol concentration simulated by Bulk-2M is close to that by SBM (Fig. 4.2) and the only aerosol sink is CCN activation in those simulations, the activation parameterization by Abdul-Razzak et al. (1998) in Bulk-2M yields a similar performance with the calculation based on the Köhler theory in the SBM. Our offline tests of both nucleation schemes further indicate that the activation parameterization by Abdul-Razzak et al. (1998) exhibits

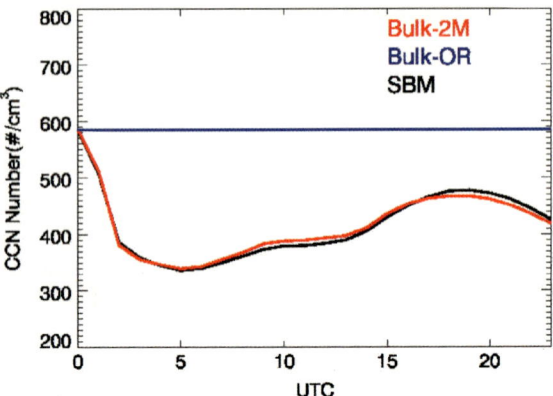

Fig. 4.2 Temporal evolution of the domain-averaged aerosol number concentration from the three simulations with SBM, Bulk-OR, and Bulk-2M in the Sc case (Reprinted from Wang et al. (2013) with permission of John Wiley and Sons)

the best agreement with the sectional approach of activation when the environmental updraft velocity is moderate (1–3 m/s). The increase of the aerosol concentration after 0500 UTC in Fig. 4.2 is explained by the fact that fresh aerosols are allowed to enter from the lateral boundaries and slowly replenish the aerosol populations within the domain. Although there is no representation of the CCN regeneration after the complete droplet evaporation in Bulk-2M, the good agreement of simulated aerosol concentrations between Bulk-2M and SBM that accounts for the CCN regeneration from the droplet evaporation (Fan et al. 2009) indicates that the CCN regeneration does not significantly contribute to the total CCN concentration in the Sc case with CCN sources set at the lateral boundaries.

4.2.2 Comparison with Field Measurements

Measurements of cloud properties by the NSF/NCAR C130 Research Aircraft (in the innermost domain from 0900 to 1300 UTC), including liquid water content (LWC) from a PMS Hot Wire Liquid Water Probe (King Probe), cloud number concentration (Nc), and cloud effective radius (Rc), are employed to evaluate the performances of the different microphysical schemes. The number concentration and the effective radius of cloud droplets within a size range from 1 to 50 μm are measured by a PMS Cloud Droplet Probe (CDP).

Comparisons of the vertical distribution of cloud properties from observation and simulated by the three microphysical schemes, SBM, Bulk-OR, and Bulk-2M, are summarized in Fig. 4.3. All three schemes predict similar LWC profiles, i.e., increasing from 500 to 1300 m. The observational data lie within the uncertainty range of the simulated profiles of LWC, indicating that the three microphysical schemes simulate LWC reasonably well. In particular, SBM has a better agreement with the field measurement than to the bulk microphysics simulations. SBM reproduces the peak of the observed LWC at about 1300 m, while the bulk simulations predict smaller peaks at lower altitudes. There are noticeable differences in *Nc* predicted (2nd column of Fig. 4.3) among SBM, Bulk-2M, and Bulk-OR. SBM

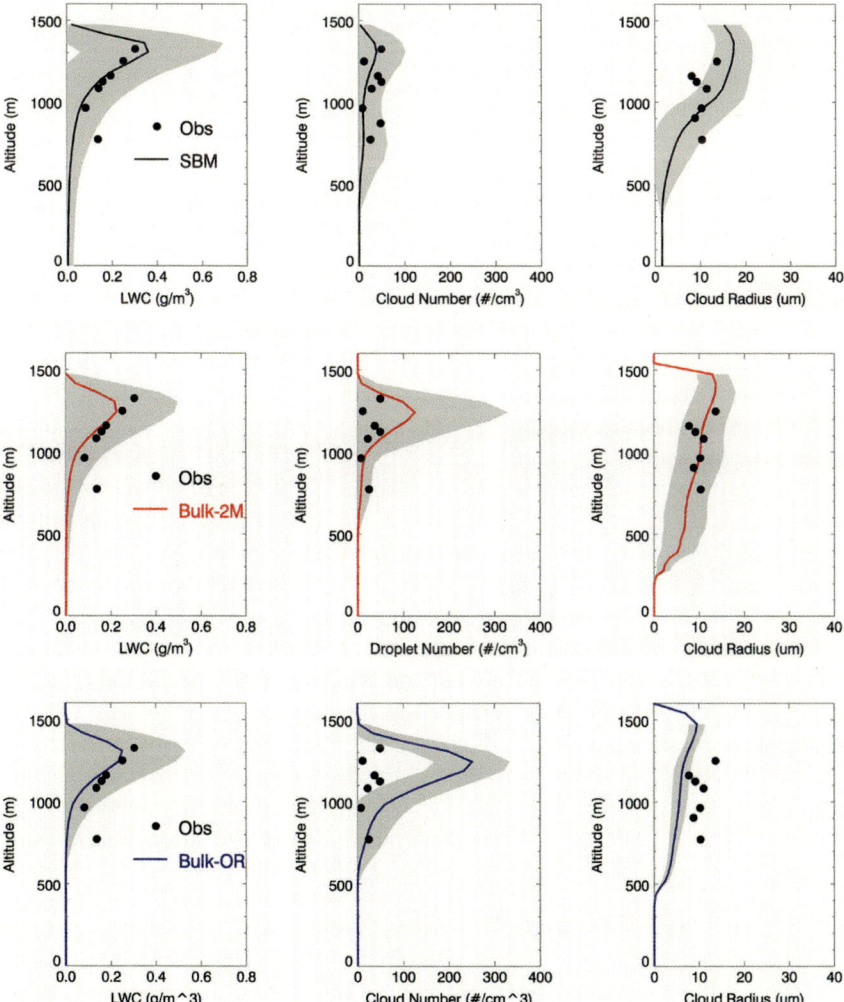

Fig. 4.3 Comparison of the vertical profiles of cloud microphysical properties from the three simulations SBM, Bulk-OR and Bulk-2M with the C130 aircraft measurements. The *first, second,* and *third columns* present the liquid water content, cloud droplet number concentration, and cloud droplet effective radius, respectively. The *black dots* denote the mean values of observations at given heights within ±25 m. *Shading areas* denote the standard derivation of the sampling data over the flight track (Reprinted from Wang et al. (2013) with permission of John Wiley and Sons)

exhibits the best performance, with the simulated Nc reproducing well with the observed values. Most of observed values are far away from the uncertainty range of Bulk-OR, which overestimates the total Nc by a factor of five. The revised scheme Bulk-2M, in which the aerosol number and mass are prognostic, shows a significant improvement in the simulation of Nc, and the observed values lie within the uncertainty range predicted by Bulk-2M. Because of the dramatic overestimation of Nc, Rc in Bulk-OR is largely underestimated (3rd column of Fig. 4.3), while Bulk-2M exhibits the best performance among the three schemes to reproduce the observed Rc profile. The good agreement between the SBM simulation and in situ measurements indicates that SBM can be used as the benchmark simulation to evaluate the performances of the bulk microphysics.

The time series of the liquid water path (LWP), height of cloud base, cloud thickness, and cloud radar reflectivity from the airborne measurements are compared with the simulations of Bulk-2M, Bulk-OR and SBM. The airborne measurements show considerable spatial variations in the cloud properties, reflecting a complex three-dimensional structure for Sc. Figure 4.4a presents a comparison of the LWP measured by water vapor radiometer on the C130 aircraft with model simulated LWP. The SBM results show the best agreement with the temporal variation of LWP long the flight path, while Bulk-OR and Bulk-2M consistently underpredict LWP. There is no significant difference between the LWP simulated by Bulk-OR and Bulk-2M, mainly because the same saturation adjustment strategy is employed for the liquid water formation. The cloud base height and cloud thickness were measured by the Wyoming cloud radar and lidar equipped on the C130 Aircraft. Figure 4.4b shows all three simulations predict the cloud base heights around 1 km, consistent with the observations. However, the SBM and bulk schemes show a large discrepancy in simulating cloud thickness. Figure 4.4c shows that only SBM reproduces the temporal evolution of the observed cloud thickness. Both Bulk-OR and Bulk-2M predict much thinner clouds during most of the simulation period. The lower cloud depth and lower LWC explain the lower LWP in Bulk-OR and Bulk-2M. Since there is no significant improvement of LWP and cloud thickness in Bulk-2M compared to Bulk-OR, it is possible that the simple treatment of condensation/evaporation in the bulk scheme, i.e., the saturation adjustment, may explain the discrepancies between the bulk and SBM simulations (to be discussed later). The measured reflectivity at 100 m (Z_{100}) from the Wyoming 94 GHz W-band cloud radar on the C130 Aircraft is compared with the model simulated reflectivity in Fig. 4.4d. The derivation of Z_{100} from the model simulated rainfall rate at 100 m (R_{100}) follows the empirical Z-R relationship: $Z_{100} = 57(R_{100}/24)^{1.1}$ (Comstock et al. 2004; Bretherton et al. 2010) and R_{100} is approximated by the simulated surface rainfall rate. Both SBM and Bulk-2M predict drizzle formation ($Z_{100} > 0$ dBz) during 0900–1300 UTC, as detected by the radar, while drizzle is largely shut off in Bulk-OR. SBM exhibits a better agreement on the magnitude of Z_{100} with the radar measurement than Bulk-2M, which tends to underpredict the radar reflectivity during the simulation. The difference between surface rainfall rate and R_{100} can partially explain the underprediction of reflectivity in all the model simulations.

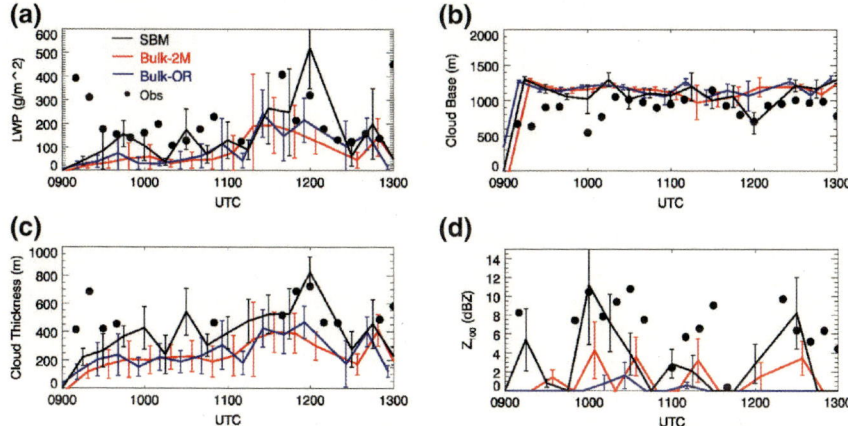

Fig. 4.4 Comparison of time series of **a** LWP, **b** cloud base height, **c** cloud thickness, and **d** radar reflectivity at 100 m from the three simulations SBM, Bulk-OR, and Bulk-2M with the C130 aircraft measurements. The *error bar* denotes the standard derivation of the sampling data over the flight track region at each altitude (Reprinted from Wang et al. (2013) with permission of John Wiley and Sons)

4.2.3 Effects on the Cloud Properties

A comparative analysis of the domain-averaged (including both cloudy and non-cloudy grids) properties of clouds and aerosols is carried out to evaluate the changes implemented to the bulk microphysics. Figure 4.5a shows that, in contrast to the large overprediction of Nc in Bulk-OR, Bulk-2M predicts Nc much closer to that simulated by SBM, because of the significantly reduced droplet nucleation rate from the prognostic treatment of the aerosol size spectrum. The averaged Nc over the cloudy-only grids in SBM, Bulk-2M and Bulk-OR is about 50, 150, and 500/cm³, respectively. Since the relatively large droplet size in the maritime cloud is favorable for the collision/coalescence process between the cloud droplets and raindrops and there is significant drizzle precipitation in this case, cloud droplet concentration is reduced significantly after the activation process. The temporal evolution of Qc averaged over the domain shows that both SBM and the bulk schemes predict similar diurnal cycles, which are regulated by the absorption of solar radiation and the consequent suppression of the total longwave radiative cooling and circulations during the daytime (Wood 2012). The comparison of Qc between the different schemes shows that SBM predicts a larger Qc than the bulk schemes. Since the cloud top radiative cooling driven by the LWC is the primary cause of the Sc formation (Wood 2012; Morrison et al. 2011), the less cloud top cooling for the lower Qc in Bulk leads to the weaker turbulence and vertical mixing. Such a feedback from the radiation-dynamics interaction further contributes to the less Qc in Bulk than in SBM. The vertical condensation/evaporation rate profiles reveal that even though the condensation rates near the cloud top are comparable between the three simulations, the peak droplet evaporation rates in the bulk schemes are much higher. The disparity in the condensation/evaporation rate between SBM and bulk schemes can be linked to the differences of the condensation/evaporation parameterization, as shown in comparative studies between the saturation adjustment in the bulk schemes and the explicit calculation of the condensation/evaporation rate with the predicted supersaturation in SBM, to be discussed in the next section.

Since Rc is significantly smaller in Bulk-OR, the collision/coalescence processes between cloud droplets are suppressed to inhibit formation of raindrops. By improving the CCN activation processes, the autoconversion efficiency in Bulk-2M is enhanced by a few orders of magnitude, resulting in a similar raindrop number concentration (Nr) as SBM (Fig. 4.5c). On the other hand, a much smaller auto-conversion rate in Bulk-OR results in significantly reduced Nr, which is only about one percent of Nr predicted by SBM. A striking difference in the simulated rain-water content (Qr) between Bulk-OR and SBM is evident in Fig. 4.5d. The amount of domain-averaged Qr from Bulk-OR is nearly negligible compared to the large amount of rainwater in SBM. Consequently, drizzle precipitation is completely inhibited in Bulk-OR, largely because of the presence of numerous but too small cloud droplets and negligible formation of raindrops. With the prognostic treatment of CCN in Bulk-2M, Nc is significantly reduced and Rc is much larger, leading to a higher autoconversion efficiency that produces much more raindrops than Bulk-OR. Therefore, more surface precipitation is produced in Bulk-2M (Fig. 4.5e). A comparison of the updraft velocities in the core grids between the three micro-physical schemes in Fig. 4.5f shows that the updraft in SBM is slightly stronger

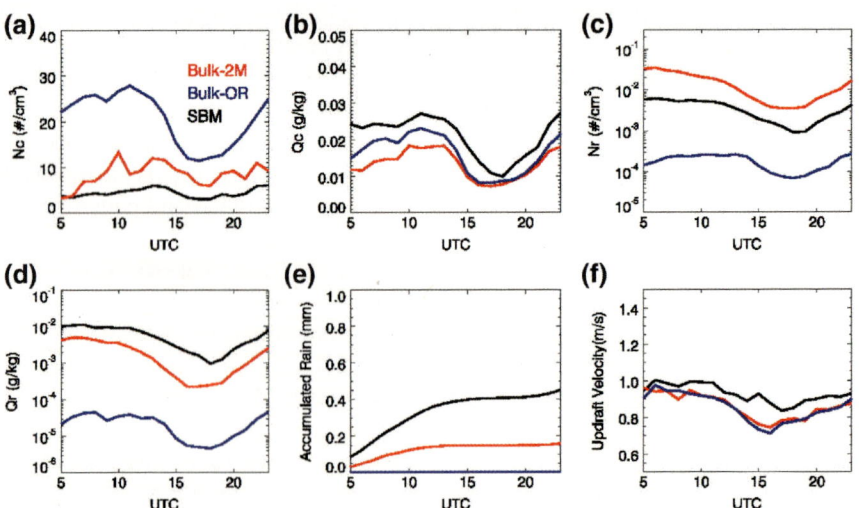

Fig. 4.5 Temporal evolution of the domain-averaged **a** cloud droplet number concentration, **b** cloud mass mixing ratio, **c** raindrop number concentration, **d** rain mass mixing ratio, **e** accumulated rainfall, and **f** core-area updraft velocity from the three simulations (Reprinted from Wang et al. (2013) with permission of John Wiley and Sons)

than the updrafts in the Bulk simulations. This result is consistent with the higher Qc in SBM (Fig. 4.5b), because a stronger updraft is more favorable for cloud formation.

4.3   Effects of Diffusional Growth Parameterizations

Note that the improved scheme Bulk-2M still predicts significantly lower Qc than SBM. To determine the cause for this difference, sensitivity experiments of a short period (3 h) are conducted using SBM and Bulk-2M, in which all the other microphysical processes, such as collision/coalescence and sedimentation processes of hydrometeors, are turned off except for the droplet nucleation and diffusion growth. The radiation processes are also turned off to preclude any feedback from the radiative cooling of the cloud top on the dynamics. The sensitivity simulations with different microphysics start from the same initial condition at 0000 UTC, Oct 28. During the first hour of the simulations (0000–0100 UTC), Qc over cloudy points remains higher in the SBM simulation than the Bulk-2M simulation after the stratocumulus clouds form within the boundary layer due to the abundant vapor supply from the ocean, as shown in Fig. 4.6a. Since the droplet nucleation rate is similar between the SBM and Bulk-2M simulations (Fig. 4.6b), the condensation/ evaporation processes impose the major influence on the liquid water budget. Figure 4.6c shows that the condensation rates, averaged over the grid points in presence of cloud droplets, are lower in the Bulk-2M simulation that utilize the saturation adjustment, compared to SBM in which explicit supersaturation is calculated for the diffusion growth. Therefore, in Sc case, the lower Qc in the bulk simulations is mainly attributed to the saturation adjustment approach employed for the condensation/evaporation processes. The updraft velocity is shown to be higher in SBM than that in Bulk-2M (Fig. 4.6d) during the first-hour simulation, indicating that the latent heat release from water vapor condensation feeds to the cloud dynamics. Through the comparison of the last 2-h simulation (0100–0300 UTC) and the first-hour simulation, it is evident that the discrepancies of Qc, condensation rate and updraft velocity between SBM and Bulk-2M simulations get significantly enlarged (Fig. 4.6e–h), supporting that there are efficient feedbacks between the cloud dynamics and microphysics. The unrealistic high Qc near the surface during the last 2-h simulation is attributed to the large updraft at low levels (Fig. 4.6h), since all sedimentation processes are turned off in this sensitive experiment. Note that the cloud droplets and rain drops share one size spectrum with an empirical cutoff size of 40 μm in SBM, in contrast to the bulk schemes that use two separate spectra for cloud and rain. These inherent differences between SBM and bulk schemes may also contribute to the different Qc and Qr in the simulations.

As a result, the treatment of the diffusion growth may need further improvements in simulating the mass content of hydrometeors and precipitation using the bulk schemes. On the other hand, the explicit calculation of the diffusional growth from the predicted supersaturation adopted in SBM may be impractical for regional or

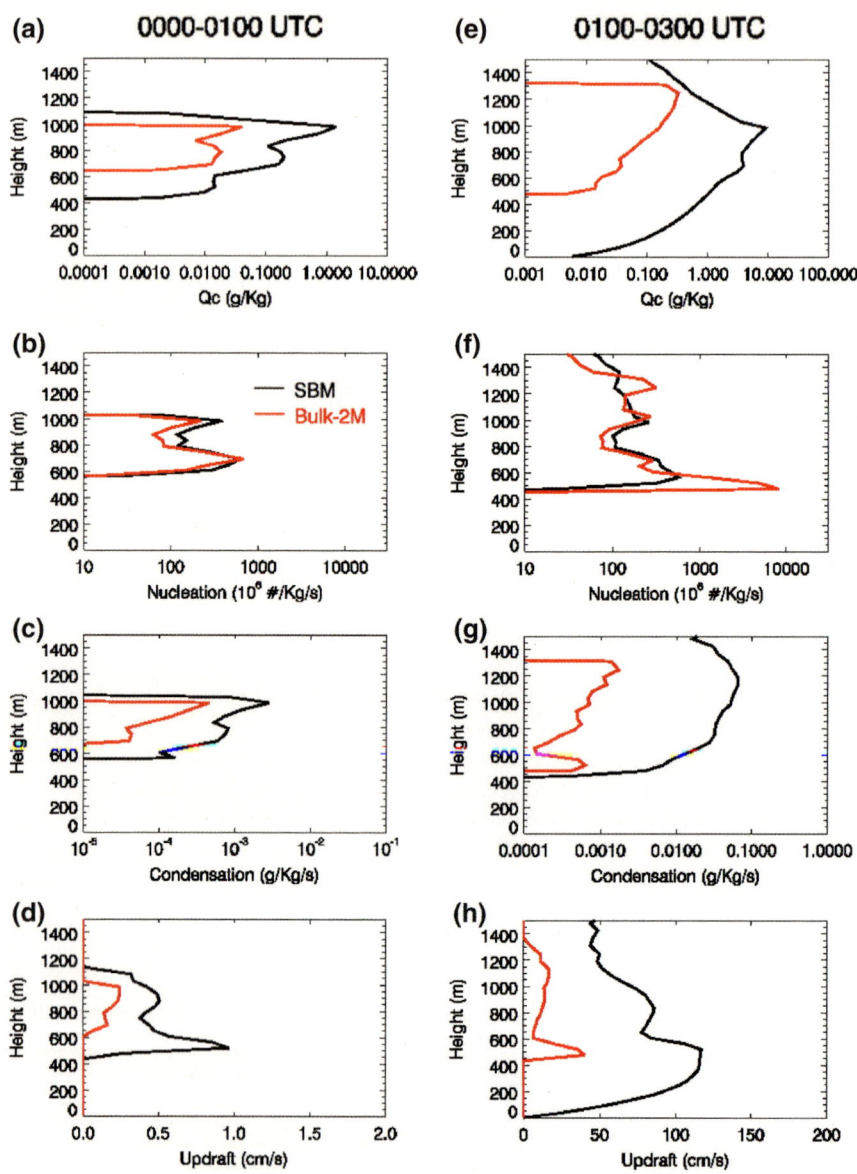

Fig. 4.6 Vertical profiles of cloud properties on cloud points from 0000 to 0100 UTC and from 0100 to 0300 UTC in the 3 h simulations using SBM and Bulk-2M without collision/sedimentation processes and radiation scheme (Reprinted from Wang et al. (2013) with permission of John Wiley and Sons)

global climate models, because of coarse resolutions. With the advances in computation power, future regional/global simulations at the cloud-resolving scales using the bulk schemes with explicit calculation of diffusional growth, such as the scheme of Li et al. (2008), may achieve better cloud simulations and more accurate assessment of AIE.

4.4   Effects of Autoconversion Parameterizations

As noted above, Qr and precipitation predicted by Bulk-2M are much lower than those by SBM. In addition to the contribution of lower Qc in Bulk-2M, there are several other factors relevant to the budget of Qr in this warm cloud case, including autoconversion, collection of cloud droplets by raindrops, and evaporation and sedimentation of raindrops. In the cloud layers, the simulated accretion rate of cloud droplets by raindrops and the evaporation rate of raindrops are on the same order of magnitude but exert opposite impacts on the budget of Qr. Therefore, the auto-conversion rate is crucial to the determination of Qr. To examine the effects of the autoconversion schemes on Qr and the drizzle precipitation, four different auto-conversion schemes are compared in the sensitivity experiments: the two-moment parameterization developed by Khairoutdinov and Kogan (2000) (hereafter referred to as KK2000) through fitting the results of large eddy simulations of marine boundary layer clouds, the two-moment parameterization developed by Seifert and Beheng (2001) (hereafter referred to as SB2001) from the stochastic collection equations, the new Kessler-type scheme developed by Liu and Daum (2004), Liu et al. (2004) (hereafter referred to as LD2004) with incorporating the relative dis-persion of the cloud droplet size distribution, and the parameterization developed by Franklin (2008), Franklin et al. (2007) (hereafter referred to as F2008) which considers the effect of turbulence on the collisions and coalescences between liquid drops.

The embryonic raindrop radius is an adjustable parameter in the autoconversion schemes and it is difficult to gauge this quantity by in situ measurements since it only describes the size of the raindrop at the newly-formed stage. In addition to the original radius of 25 μm for embryonic raindrops used in the KK2000 autocon-version scheme, an alternative value of 40 μm is also considered for consistency with the other schemes in the sensitivity tests. A comparison of Nr between the original radius of 25 μm (Fig. 4.5c) and the modified radius of 40 μm (Fig. 4.7c) for embryonic raindrops in KK2000 shows that the scheme with the modified radius predicts much lower Nr but larger raindrops, consistent with SBM simulation. This suggests that the size of 40 μm is appropriate to reflect raindrop formation in stratocumulus clouds.

Figure 4.7c, d indicate that the different autoconversion parameterizations pro-duce distinct effects on the production rates of raindrops under the same model configurations. LD2004 and F2008 predict higher autoconversion rates than the other schemes. With more efficient cloud-to-rain transformation, Qc and Nc are reduced, while Qr and Nr are elevated in LD2004 and F2008. Compared to the SBM results, the simulations of Nc, Qr, and Nr in LD2004 and F2008 are sig-nificantly improved because of the increased autoconversion rate. In F2008, the

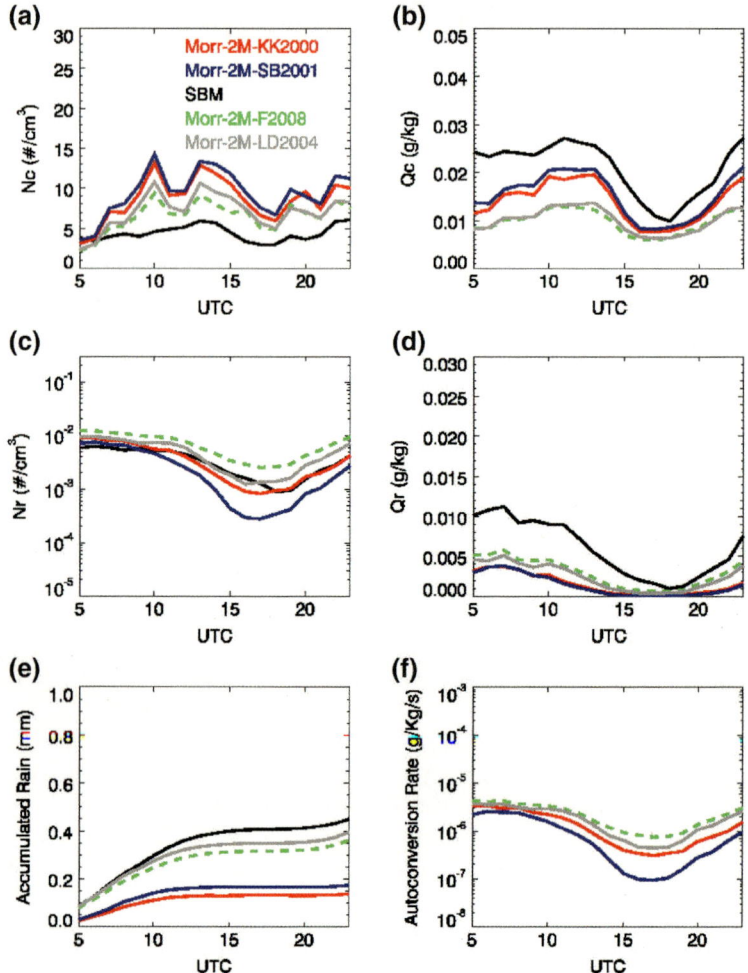

Fig. 4.7 Temporal evolution of the domain-averaged cloud properties from four bulk simulations using the different autoconversion schemes and SBM (Reprinted from Wang et al. (2013) with permission of John Wiley and Sons)

eddy dissipation rate of the turbulent kinetic energy is as high as 800 cm² s⁻³, indicating that turbulence contributes significantly to the efficient collision/coalescence processes. Furthermore, drizzle reaching to the surface is enhanced by two times in these two schemes, which is also close to the SBM results (Fig. 4.7e). Therefore, for maritime stratocumulus clouds, the autoconversion schemes of LD2004 with the dispersion factor considered and F2008 with the turbulence effects incorporated show much better performance than those of the other schemes. Li et al. (2008) also suggested that the precipitation simulated in a cumulus cloud case exhibited a large variation among the different autoconversion parameterizations,

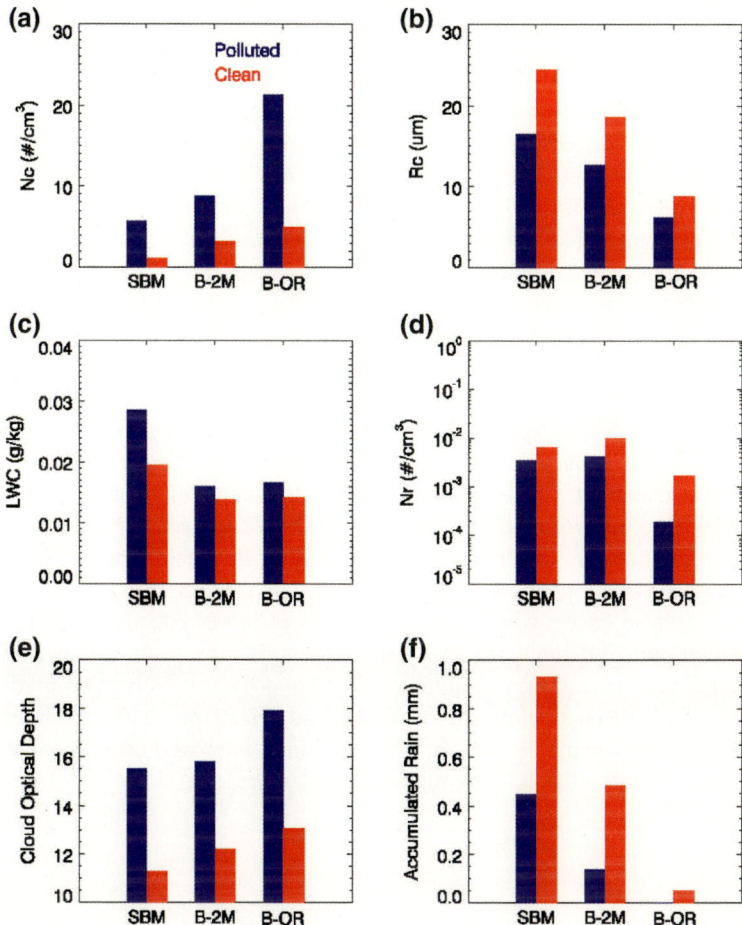

Fig. 4.8 Comparisons of the domain-averaged cloud properties from simulations with SBM, Bulk-OR, and Bulk-2M under clean and polluted (control) aerosol conditions (Reprinted from Wang et al. (2013) with permission of John Wiley and Sons)

and among them LD2004 was a more physically meaningful parameterization than the other autoconversion schemes.

4.5 Effects of Aerosol Representation on AIE

To investigate the effects of aerosol loading on stratocumulus clouds using the bulk microphysical scheme with/without the prognostic aerosol representation, simulations are performed with initialization for different aerosol loadings. As discussed above, the initial aerosol number 588 particles/cm^3 and mass concentration of

2.4×10^{-11} g/cm^3 in the control run likely represent the maritime polluted conditions under of the influence of continental outflows. A maritime clean scenario containing one sixth of the aerosol amount in the polluted case is also considered, i.e., with the aerosol number and mass concentration of 98 particles/cm^3 and 4×10^{-12} g/cm^3, respectively. Figure 4.8 shows that, in both polluted and clean cases, Bulk-2M predicts much closer Nc, Nr, LWC, Rc, and accumulated precipitation to SBM than Bulk-OR, indicating that Bulk-2M consistently outperforms Bulk-OR independent of the initial aerosol concentrations. All three simulations predict elevated cloud droplet concentration, enhanced LWC, reduced raindrops, and suppressed drizzle precipitation in the polluted case, when the aerosol concentration is six times higher. Since the saturation adjustment does not account for the influence of the droplet size distribution on the diffusional growth, the enhanced LWC in the polluted case simulated by the bulk schemes can be attributed to the smaller terminal falling speed of the size-reduced droplets and the consequent impacts on the dynamical conditions in the polluted case compared to the clean case. Those effects of sulfate aerosols on the stratocumulus cloud are consistent with previous observations (L'Ecuyer et al. 2009; Zheng et al. 2011). The cloud optical depth is mainly determined by LWP and Rc for liquid clouds. Figure 4.8e shows that Bulk-2M predicts a cloud optical depth closer to SBM in both the control and clean cases, while Bulk-OR produces more reflective clouds with a larger cloud optical depth, particularly in the control case, leading to much larger aerosol indirect forcing. Hence, although SBM and bulk microphysics predict the same sign for AIE on shallow stratocumulus and drizzle precipitation, the magnitudes of radiative forcing of AIE can be significantly improved using more realistic aerosol representation in the bulk microphysical schemes.

4.6 Summary

In this study, we have modified the widely adopted double-moment bulk cloud microphysics in the WRF model implemented by Morrison et al. (2005), with the purpose of providing a better bulk microphysical scheme for simulating aerosol effects in regional and global models. The main improvement in the modified bulk scheme (Bulk-2M) lies in the more realistic representation of aerosols by implementing prognostic aerosol mass and number concentrations into the original bulk scheme (Bulk-OR), in which aerosols are predescribed. In Bulk-2M, the activation removal and dynamical transport of aerosol particles are considered and the aerosol sources are defined at the boundaries to replenish aerosol loading and prevent dilution of aerosol concentrations by the inflow air mass. The impacts of the parameterizations of the droplet diffusional growth and autoconversion as well as the selection of embryonic raindrops radius on the performance of the bulk microphysical scheme with the prognostic double-moment aerosol representation by comparing the Bulk-2M results with those from SBM simulations are evaluated. Model simulations are conducted for maritime warm Sc regimes over southeast Pacific Ocean from the VOCALS project. The comparison between model results and atmospheric field observations reveals that SBM exhibits a better agreement with the observations due to more realistic representations in many microphysical processes, such as droplet diffusional growth and rain evaporation (Li et al. 2009; Khain et al. 2009). In this present work, the SBM scheme is considered as a benchmark to evaluate the performance of bulk microphysical schemes.

Because of the absence of aerosol advection and removal mechanisms, the unrealistic aerosol concentration temporal evolution simulated by Bulk-OR is responsible for the distinct cloud properties (i.e., Nc, Qc, Rc) with Bulk-OR from those with SBM simulations and observations, although all three model simulations (Bulk-OR, Bulk-2M, and SBM) are initiated with the same initial aerosol concentration on the basis of field observations. The simulations of cloud properties in Bulk-2M, particularly the cloud droplet number concentration and droplet size, have been significantly improved. Because of numerous small droplets simulated by Bulk-OR, the conversion of cloud droplets to raindrops is very inefficient, and the rainwater and drizzle formation is largely inhibited. Bulk-2M predicts close simulations to SBM in terms of the realistic rainwater and drizzle formation. With an equivalent aerosol budget to the SBM, Bulk-2M is further examined for parameterizations in the bulk microphysics, on the basis of comparing with the results from the SBM that can be considered as a benchmark. Taking the advantage of the prognostic aerosol approach, the observed aerosol/CCN size distributions can be employed to properly initiate the simulations, avoiding tuning the aerosol/droplet concentrations. The prognostic aerosol approach will also be essential, when the advective/convective transports of aerosols are efficient for convective cloud regimes. We will incorporate CCN regeneration scheme and wet scavenging to fully represent CCN budget in the future study.

There still exist large discrepancies in the predicted cloud water and rainwater between Bulk-2M and SBM, and Bulk-2M predicts significantly lower cloud water and rainwater. Sensitivity experiments indicate that the vapor saturation adjustment employed for the diffusion growth in the bulk schemes mainly contributes to the lower cloud water due to a low condensation rate and a high evaporation rate compared to SBM, in which diffusional growth is explicitly calculated based on supersaturation. Since explicit calculation of the diffusional growth employed in SBM is impractical for regional and global models because of coarse resolutions, bulk schemes with explicit calculation of diffusional growth such as those of Li et al. (2008), Lebo et al. (2012) may be adapted for further improvements in the cloud-resolving simulations.

The low rainwater amount in Bulk-OR is shown to be closely related to the KK2000 autoconversion parameterization. Our sensitivity tests indicate that LD2004, in which the dispersion factor is explicitly incorporated, and F2008 with the consideration of turbulence effect exhibit good performances in terms of realistic simulations of cloud droplet number concentration, raindrop number and mass concentration, and drizzle precipitation for stratocumulus clouds. Through improvement in the simulations of Nr and the effective rain radius in the Sc case, the size of 40 μm for embryo raindrops is demonstrated to be more realistic than that used in the original KK2000 scheme (25 μm).

Sensitivity modeling experiments have been performed to evaluate the responses of increasing aerosols in model simulations for Sc cloud regimes. Bulk-2M simulates the magnitudes of aerosol effects on the cloud droplet number and droplet size, cloud optical depth, and precipitation close to those by SBM. In summary, our results demonstrate that better cloud simulations and more accurate assessment of aerosol indirect effects can be achieved in regional and global models by improving aerosol representation, more sophisticated calculation of the diffusion growth, and more realistic autoconversion parameterizations in the bulk microphysics schemes.

References

Abdul-Razzak H, Ghan SJ, Rivera-Carpio C (1998) A parameterization of aerosol activation-1. Single aerosol type. J Geophys Res 103(D6):6123–6131

Bretherton CS, Wood R, George RC, Leon D, Allen G, Zheng X (2010) Southeast Pacific stratocumulus clouds, precipitation and boundary layer structure sampled along 20° S during VOCALS-REx. Atmos Chem Phys 10(21):10639–10654

Comstock KK, Wood R, Yuter SE, Bretherton CS (2004) Reflectivity and rain rate in and below drizzling stratocumulus. Q J R Meteorol Soc 130(603):2891–2918

Fan JW, Ovtchinnikov M, Comstock JM, McFarlane SA, Khain A (2009) Ice formation in Arctic mixed-phase clouds: insights from a 3-D cloud-resolving model with size-resolved aerosol and cloud microphysics. J Geophys Res 114. doi:10.1029/2008JD010782

Franklin CN (2008) A warm rain microphysics parameterization that includes the effect of turbulence. J Atmos Sci 65(6):1795–1816

Franklin CN, Vaillancourt PA, Yau MK (2007) Statistics and parameterizations of the effect of turbulence on the geometric collision kernel of cloud droplets. J Atmos Sci 64(3):938–954

Hong S-Y, Noh Y, Dudhia J (2006) A new vertical diffusion package with an explicit treatment of entrainment processes. Mon Weather Rev 134:2318–2341

Khain AP, Leung LR, Lynn B, Ghan S (2009) Effects of aerosols on the dynamics and microphysics of squall lines simulated by spectral bin and bulk parameterization schemes. J Geophys Res 114:D22

Khairoutdinov M, Kogan Y (2000) A new cloud physics parameterization in a large-eddy simulation model of marine stratocumulus. Mon Weather Rev 128(1):229–243

L'Ecuyer TS, Berg W, Haynes J, Lebsock M, Takemura T (2009) Global observations of aerosol impacts on precipitation occurrence in warm maritime clouds. J Geophys Res 114:D9

Lebo ZJ, Morrison H, Seinfeld JH (2012) Are simulated aerosol-induced effects on deep convective clouds strongly dependent on saturation adjustment? Atmos Chem Phys 12 (20):9941–9964

Li G, Wang Y, Zhang R (2008) Implementation of a two-moment bulk microphysics scheme to the WRF model to investigate aerosol-cloud interaction. J Geophys Res 113(D15):D15211

Li X, Tao W-K, Khain AP, Simpson J, Johnson DE (2009) Sensitivity of a cloud-resolving model to bulk and explicit bin microphysical schemes. Part II: Cloud microphysics and storm dynamics interactions. J Atmos Sci 66(1):22–40

Liu YG, Daum PH (2004) Parameterization of the autoconversion process. Part I: Analytical formulation of the Kessler-type parameterizations. J Atmos Sci 61(13):1539–1548

Liu YG, Daum PH, McGraw R (2004) An analytical expression for predicting the critical radius in the autoconversion parameterization. Geophys Res Letts 31(6)

Morrison H, Curry JA, Khvorostyanov VI (2005) A new double-moment microphysics parameterization for application in cloud and climate models. Part I: Description. J Atmos Sci 62(6):1665–1677

Morrison H et al (2011) Intercomparison of cloud model simulations of Arctic mixed-phase boundary layer clouds observed during SHEBA/FIRE-ACE. J Adv Model Earth Syst 3: M06003. doi:10.1029/2011MS000066

Seifert A, Beheng KD (2001) A double-moment parameterization for simulating autoconversion, accretion and selfcollection. Atmos Res 59:265–281

Wang H, Feingold G, Wood R, Kazil J (2010) Modelling microphysical and meteorological controls on precipitation and cloud cellular structures in Southeast Pacific stratocumulus. Atmos Chem Phys 10(13):6347–6362

Wang Y, Fan J, Zhang R, Leung R, Franklin C (2013) Improving bulk microphysics parameterizations in simulations of aerosol indirect effects. J Geophys Res 118:1–19

Wood R (2012) Stratocumulus clouds. Mon Weather Rev 140(8):2373–2423

Yang Q, Gustafson JWI, Fast JD, Wang H, Easter RC, Morrison H, Lee YN, Chapman EG, Spak SN, Mena-Carrasco MA (2011) Assessing regional scale predictions of aerosols, marine stratocumulus, and their interactions during VOCALS-REx using WRF-Chem. Atmos Chem Phys 11(23):11951–11975

Zhao C et al (2006) Aircraft measurements of cloud droplet spectral dispersion and implications for indirect aerosol radiative forcing. Geophys Res Lett 33:L16809

Zheng X et al (2011) Observations of the boundary layer, cloud, and aerosol variability in the southeast Pacific near-coastal marine stratocumulus during VOCALS-REx. Atmos Chem Phys 11(18):9943–9959

Chapter 5
Impacts of Asian Pollution Outflows on the Pacific Storm Track

Abstract Increasing levels of particulate matter pollutants over the continents and associated pollution outflows have raised considerable concerns because of their potential impacts on the Pacific storm track and the regional climate. Inspired by the previous research work in our group (Zhang et al. 2007; Li et al. 2008a), which investigated the interannual trend of deep convective clouds and precipitation over the North Pacific from the observational dataset, I kept seeking the observational evidences and theoretical basis for the impacts of Asian pollution outflows on the storm track. In this study, we have derived and identified the interannual variation of Northwest Pacific storm track intensity on the basis of reanalysis datasets. Various modeling approaches have been utilized to tackle the multiscale nature of the interaction between aerosols and large-scale circulations. Seasonal (2 month) simulations using a CR-WRF model with a two-moment bulk microphysics have been conducted to examine the aerosol effects on the regional climate over the Pacific storm track. Subsequently, the anomalies of the diabatic heating rates produced by the Asian pollution outflow from the CR-WRF simulations are prescribed in the NACR Community Atmosphere Model (CAM5) as the aerosol forcing terms. Simulations of three winters using CAM5 with and without the derived aerosol forcings have been performed. In addition, the comparisons to the results from the multiscale aerosol-climate modeling framework have been conducted to further validate the responses of the Pacific storm track to the different aerosol forcings associated with elevated pollution levels.

Keywords Multiscale modeling · Asian pollution · Pacific storm track

5.1 Observational Evidences

Different with the previous examination of the storm track's signature, such as deep convective cloud (Zhang et al. 2007) and precipitation (Li et al. 2008a), the variation of the Pacific storm track intensity has been directly assessed using the

© Springer-Verlag Berlin Heidelberg 2015   55
Y. Wang, *Aerosol-Cloud Interactions from Urban, Regional, to Global Scales*,
Springer Theses, DOI 10.1007/978-3-662-47175-3_5

National Center for Environmental Predictions and Department of Energy (NCEP/DOE) Reanalysis II. This dataset is an improved version of the NCEP Reanalysis I model that fixed errors and updated parameterizations of physical processes. The 6 h fields at 17 pressure levels are available with a T62 resolution ($\sim$210 km) from 1979. The European Centre for Medium-Range Weather Forecasts (ECMWF) Interim Reanalysis Data is later analyzed to validate the results from NCEP Reanalysis data.

Two indices derived from reanalysis data are used to characterize storm track intensity: the 2–8 day band-pass-filtered transient eddy meridional heat flux at 850 hPa (EMHF $= \overline{v'_{850}T'_{850}}$) and the 2–8 day band-pass-filtered transient eddy meridional wind variance at 300 hPa (EMWV $= \overline{v'^2_{300}}$), where v' and T' denote the eddy meridional velocity and eddy temperature departing from the mean seasonal cycles. The 2–8 day band-pass filter is used with respect to the fact that the time periods of storms over the Pacific are typically shorter than 1 week. As an important component of the total poleward heat transport, EMHF is typically used to quantify the "storminess" in the middle latitude (Nakamura et al. 2002). At 300 hPa, which is referred as the jet stream level, the meridional wind variance is associated with the storm track activities (Trenberth 1997).

We study the EMHF and EMWV of January and February over the NW Pacific region (20°–60° N, 120° E–170° W) from 1979 to 2011. The empirical orthogonal function (EOF) analysis is employed on the time series of EMHF and EMWV from the NCEP data, so the results are reduced to a set of orthogonal vectors that simply represent the pattern of variance in the spatial distribution. The leading principal component (PC1) corresponds to the simultaneous strengthening/weakening pattern of the storm tracks (Chang et al. 2002). Figure 5.1a shows the leading principal component of the heat flux over the NW Pacific along time series, which equalizes the temporal variation of the Pacific storm track intensity. The variance percentage of PC1 is 35.4 % meaning that the first mode accounts for about 35.4 % of the total variance in heat flux. The linear regression has been performed onto PC1 of heat flux and the positive slope of regression line on the heat flux implies that the overall trend of the heat flux associated with Pacific storm track has been increased during recent 28 years in spite of some fluctuations. Statistical F-test on the regression slop is performed and the confident level is greater than 90 %.

As an alternative measure of storm track intensity, the eddy meridional wind variance has been examined in this study. Similarly, EOF analysis is employed and its leading principal component is used as another index of storm track intensity. Utilizing a linear regression, the leading principal component of the wind variance associated with the Pacific storm track is found to increase by 15.5 m^2/s^2 per year since 1980, as shown in Fig. 5.1b. The variance percentage for PC1 accounts for 42 % of the total variances. The confidence level of F-test on linear regression is greater than 95 %.

Considering the fast economic growth in Asian countries, two decadal periods, 1979–1988 and 2002–2011, can be used to represent two different economic developing stages in Asian countries with different pollution levels. As shown in

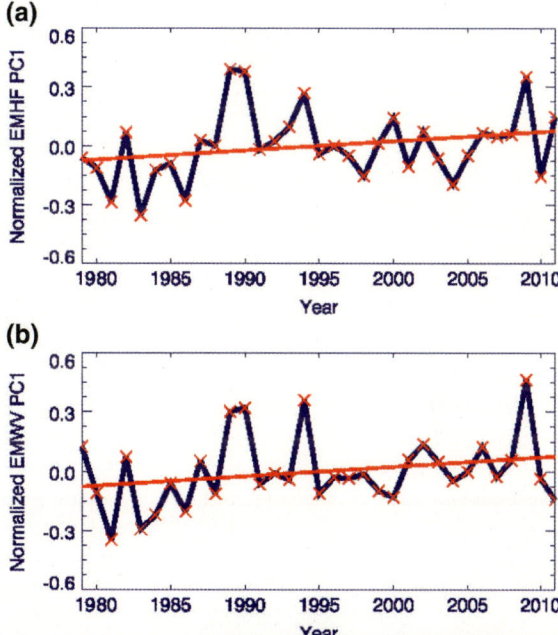

Fig. 5.1 Leading principal component of **a** 850 mb EMHF and **b** 300 mb EMWV over the northwest Pacific

the geospatial distributions of decadal mean EMHF at 850 hPa (Fig. 5.2a, b), the heat flux variances are concentrated in the storm track downstream region along 40° N latitude in the NW Pacific. The comparison of EMHF between those two decades (Fig. 5.2c) reveals that there is an evident increase of EMHF over the storm track in the decade of 2002–2011 with a more polluted condition. Averaged over the NW Pacific region, the EMHF is increased by 8.6 % in 2002–2011. The statistical significant tests support that the increases of EMHF on the most of grid points have a confident level greater than 90 %. The enhancement of the NW Pacific storm track intensity can also be identified through the examination of the EMWV spatial distribution over the NW Pacific in two periods 1979–1988 (Fig. 5.2d) and 2002–2011 (Fig. 5.2e). The difference of EMWV between the two periods (Fig. 5.2f) reveals that the recent EMWV is increased by 9.7 % compared to the low-pollution condition in early 1980s. Even though the maximal EMWV area is located in the west of 180° E longitude, the area with the EMWV enhancement is located in the NW Pacific, which is close to the Asian pollution sources and consistent with the distribution of EMHF difference between the two periods (Fig. 5.2f). We also analyze the European Center for Medium-Range Weather Forecasting (ECMWF) ERA-Interim reanalysis dataset, and the similar enhancements of EMHF and EMWV are found between two decades (Fig. 5.3). The next step is to simulate the impacts of Asian pollution outflows on the Pacific storm track.

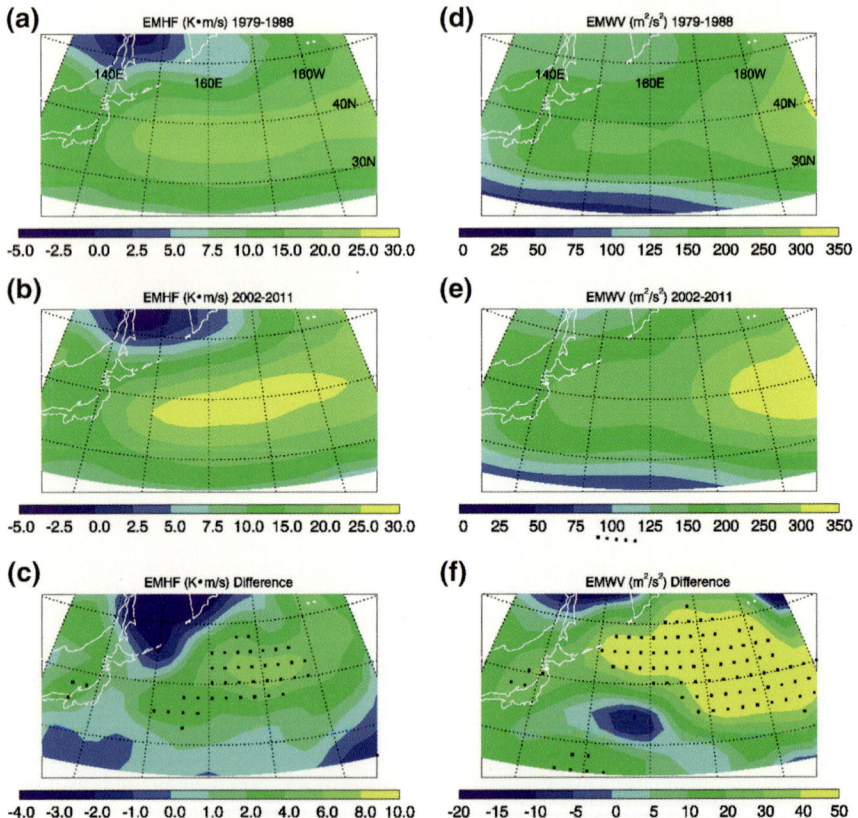

Fig. 5.2 Spatial distribution of the EMHF at 850 hPa during **a** 1979–1988, **b** 2002–2011, **c** difference between the two decades and the EMWV at 300 hPa during **d** 1979–1988, **e** 2002–2011, **f** difference between the two decades over the northwest Pacific from NECP/DOE Reanalysis II Dataset. *Dots* in (**c**) and (**f**) indicate the 90 % level of statistical significance based on the Student's t-test

5.2 A Hierarchical Modeling Approach

In this study, we introduce a hierarchical modeling approach to incorporate aerosol effects on the deep convective clouds from cloud-resolving model to the global climate model and investigate the large-scale influence of aerosols on the storm track dynamics. Considering that diabatic heating is determinant in maintaining the baroclinicity in the storm track entrance regions (Chang et al. 2002), the variation of diabatic heating induced by aerosol will be used to echo the aerosol effects from cloud-resolving model to the global climate model. As the first step, the seasonal simulations of Pacific storm track with and without the consideration of the Asian pollution outflows were carried on using the CR-WRF model and the aerosol effects were investigated in the cloud-resolving scale. The monthly mean diabatic heating

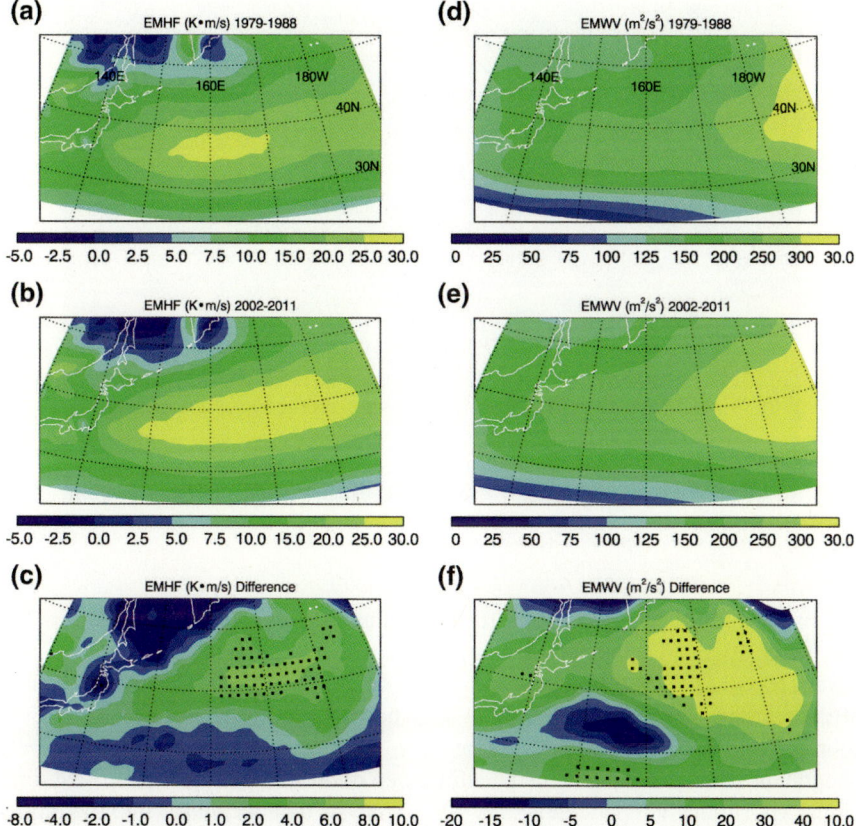

Fig. 5.3 Spatial distribution of the eddy meridional heat flux (EMHF) at 850 hPa during **a** 1979–1988, **b** 2002–2011, **c** difference between the two decades and the eddy meridional wind variance (EMWV) at 300 hPa during **d** 1979–1988, **e** 2002–2011, **f** difference between the two decades over the northwest Pacific from ECMWF ERA-Interim Dataset. *Dots* in (**c**) and (**f**) indicate the 90 % level of statistical significance based on the Student's t-test

rates were calculated and the differences between different aerosol scenarios were implemented in the NCAR CAM5 model. The storm track dynamics during three winter seasons over the North Pacific were assessed using the CAM5 model with modified diabatic heating sources.

5.2.1 Configuration of CR-WRF and Experiment Design

External forcing of WRF model is considered from the NCEP Final Global Analyses (FNL) $1° \times 1°$. The model was configured with two one-way interactive domains. To provide most coverage of the NW Pacific, the outer domain has the

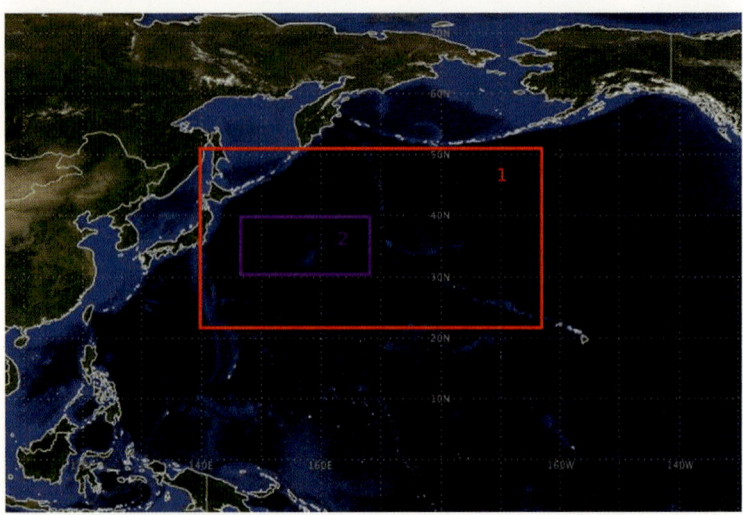

Fig. 5.4 Domains overview for CR-WRF simulations

size of 4000 km × 3000 km with a 10-km resolution in horizontal and is centered at (40° N, 170° E). The nested domain provides a finer resolution of 2.5 km and covers 1000 km × 1500 km area with the center of (36° N, 160° E). The overview of domains is shown in the Fig. 5.4. Stretched 35 Eta levels are used in vertical coordinate with about 500 m spacing. The third-order Runge–Kutta-based time integration scheme is adopted using a time step of 60 s in the simulation. Real-time global sea surface temperature (RTG_SST) analysis data are added into the model as an extra forcing at each input time step. The Dudhia and RRTM schemes have been used as shortwave and longwave radiation schemes, respectively. For the boundary layer physics scheme, we choose the YSU scheme. No cumulus cloud parameterization was used. A 2-month simulation has been performed from January 1st to February 28th, 2003. The whole simulation is composed of eight sections and each section is an 8-day run. Then the model is restarted using initial conditions from NCEP FNL data in the beginning of each section. There is a 1-day overlap between nearby sections to make sure the model spins up sufficiently.

To investigate the aerosol effect on the storm track, we conducted the sensitivities experiments with two aerosol scenarios under the same dynamic and thermodynamic conditions, the polluted case with consideration of the Asian outflow of aerosol (P-case) and the maritime clean case (M-case). Two aerosol species are considered: the polluted continental aerosols are assumed to mainly contain ammonium sulfate and maritime aerosols are assumed to be sodium chloride (Seinfeld and Pandis 2006). Monthly average ammonium sulfate data from the Model of Ozone And Related Tracers (MOZART) Version 2.2 was utilized to interpolate aerosol initial and boundary fields. MOZART (Tie et al. 2001) is a global chemical transport model and the monthly average production provides more than 80 chemical species with a

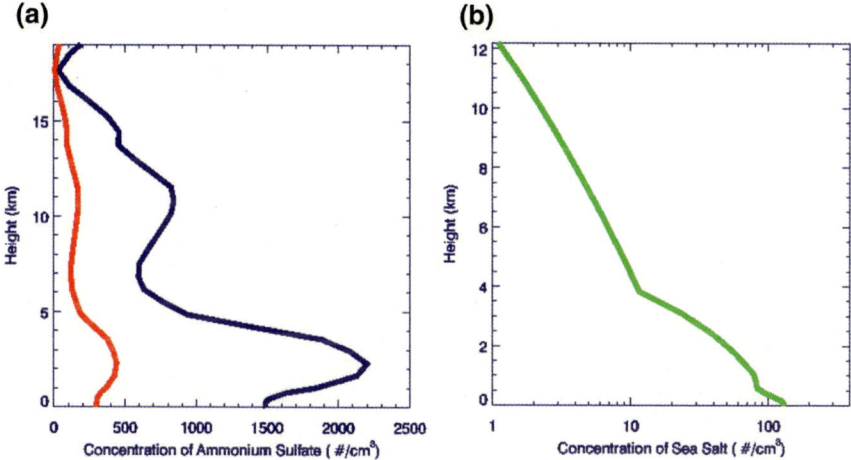

Fig. 5.5 Vertical profile of aerosol number concentration in the initial conditions of CR-WRF. **a** Number concentration profile of ammonium sulfate (cm^{-3}). *Red line* denotes M-case and *dark blue line* denotes P-case. **b** Number concentration profile of sea salt (cm^{-3})

horizontal resolution of 2.8° and 34° vertical levels. Figure 5.5a shows the vertical profiles of the ammonium sulfate number concentration. In the P-case, background ammonium sulfate data was initiated as the profile with the maximal number concentration of 2200 cm^{-3} at about 2.5 km level, while the initial profile in the M-case has about 440 cm^{-3} ammonium sulfate as the maximal value. Some aircraft measurements reported the exponentially decreasing profile of sea salt in vertical over the northwest Pacific (Zhao et al. 2006). The background sea salt is initiated as the profile with the maximal number concentration about 110 cm^{-3} at the surface for both P-case and C-case as shown in Fig. 5.5b.

A higher scalar profile of ammonium sulfate is fixed in the western boundary of the outer domain to mimic the Asian pollution outflow in the P-case. Under the aerosol horizontal turbulence diffusion scheme, the strong zonal winds over the NW Pacific transport aerosols in the entire domain. A sea-salt emission scheme considering wind velocity and relative humility over the sea surface was included in the model (Li et al. 2008a).

5.2.2 Evaluations of CR-WRF Simulations

Satellite observational data has been utilized to validate the results from CR-WRF model simulation, especially in the polluted case considering the Asian pollution outflow. We compare several model results in the P-case with observational product, to investigate the development of cyclones, cloud properties, and precipitation over the NW Pacific.

5.2.2.1 Cyclones Development

Mid-latitude cyclones are prevalent over the storm track region (Chang et al. 2002) and traditional definition of the storm track is based on the cyclone statistics. Therefore, it is important to reproduce the development of cyclones in the model simulation. In the wintertime of the NW Pacific, cyclones usually last about 1 week over the Pacific and there were about eight storm events during the 2-month period in the simulations. As 3-day snapshots of the storm event, MODIS Terra Collection 5 Level 3 daily data on January 20, 21 and 22, 2003 (Fig. 5.6d–f) provides the detailed view of what was happening over the NW pacific. On January 20, an embryo storm developed in the west and 1 day later the large-scale cyclone formed and swept the center of the NW Pacific. On January 22, the cyclone moved toward downstream of storm track and dissipated in the domain while at the same time another storm developed and took place in the upstream of the storm track. On comparison of the cloud water path between CR-WRF model result (Fig. 5.6a–c) and MODIS data(Fig. 5.6d–f), our simulations successfully reproduce features mentioned above. The size and location of the storms are also consistent between the observation and model simulations. Hence, the CR-WRF model shows a good performance in resolving the cyclone and simulating the storm events.

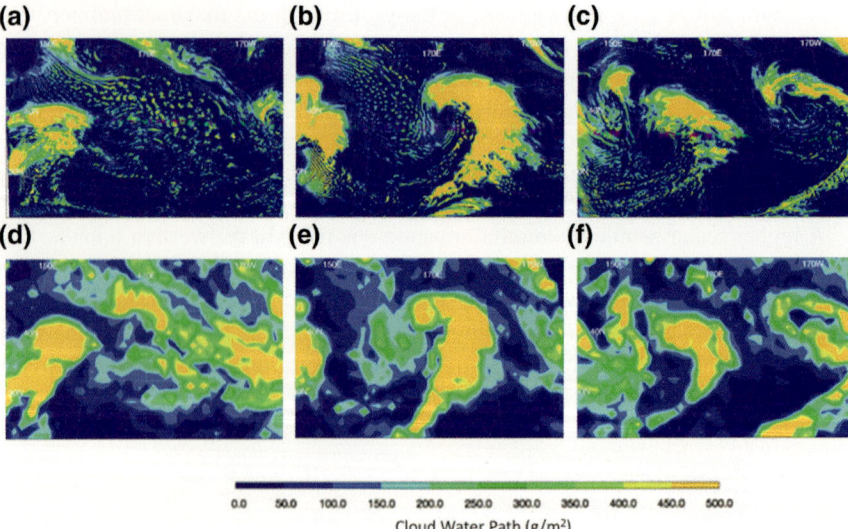

Fig. 5.6 Snapshot of storm event. **a** CR-WRF simulated cloud water path in Jan 20. **b** CR-WRF simulated cloud water path on Jan 21. **c** CR-WRF simulated cloud water path on Jan 22. **d** MODIS Terra L3 Daily cloud water path on Jan 20. **e** MODIS Terra L3 Daily cloud water path on Jan 21. **d** MODIS Terra L3 Daily cloud water path on Jan 22 (Reprinted from Wang et al. (2014a) with permission of Nature Publication Group)

5.2.2.2 Cloud Fraction

The cloud simulation from CR-WRF model result are compared with the MODIS Collection 5 Level 3 monthly data with a gridded $1° \times 1°$ global data from Terra satellite. Figure 5.7 shows the monthly statistical mean of cloud fraction in January and February 2003. The overall patterns of cloud fraction between the MODIS data (Fig. 5.7a, c) and the model results (Fig. 5.7b, d) are well consistent, and both of them depict clearly the pattern of the higher cloud fraction (>0.7) areas in the north part of the domain. Note that the domain-averaged cloud fraction over the wintertime Pacific is constantly larger than 50 %, implying that aerosol indirect effects should play a more dominant role compared to aerosol direct effect.

5.2.2.3 Precipitation

The simulated precipitation in the outer domain is compared with the $0.25° \times 0.25°$ monthly Tropical Rainfall Measuring Mission (TRMM) 3B43 data. The dataset combines satellite observation with gauge data and provides the latitude coverage from 50° N–50° S. Figure 5.8 displays the TRMM monthly accumulated rainfall in January and February over the NW pacific area. A rainfall belt across the entire domain can be identified easily. This is due to the fact that storm track downstream is characterized by a belt of local maximal precipitation across almost the entire north Pacific (Xie and Arkin 1997). The distribution patterns of simulated precipitation (Fig. 5.8b, d) are in agreement with the TRMM observation (Fig. 5.8a, c). In January's simulation, the CR-WRF model well reproduces the three centers with intensive precipitation along 35° N latitude: 150° E–160° E, 170° E–180° E, and around 170° W. For the simulation in February, the model successfully yields the rainfall peak around 35° N, 170° W and its overall magnitude of rainfall is closer to the TRMM data over the whole NW Pacific. It is worthy to note that in some grids the simulation results overestimate the amount of precipitation compared with TRMM data, while in some other grids the maximal precipitation from TRMM data is larger than that from simulation results. This discrepancy can be attributed to the bias of the model as well as observation data. On the observation side, the very limited rain gauge data is available to apply a large-scale bias adjustment over the mid-latitude ocean and the high values of root-mean-square (RMS) precipitation-error estimates over the NW Pacific are found in the RMS precipitation-error map (not shown here). On the modeling side, the limitations of numerical scheme and relatively coarse resolution (10 km) of the outer domain might lead to differences between the simulation and observation.

5.2.3 Sensitivity Study and Derived Aerosol Forcings

In the wintertime, the NW Pacific region is characterized by migrating convective clouds and trailing stratiform clouds. The total frequency of clouds at all levels is

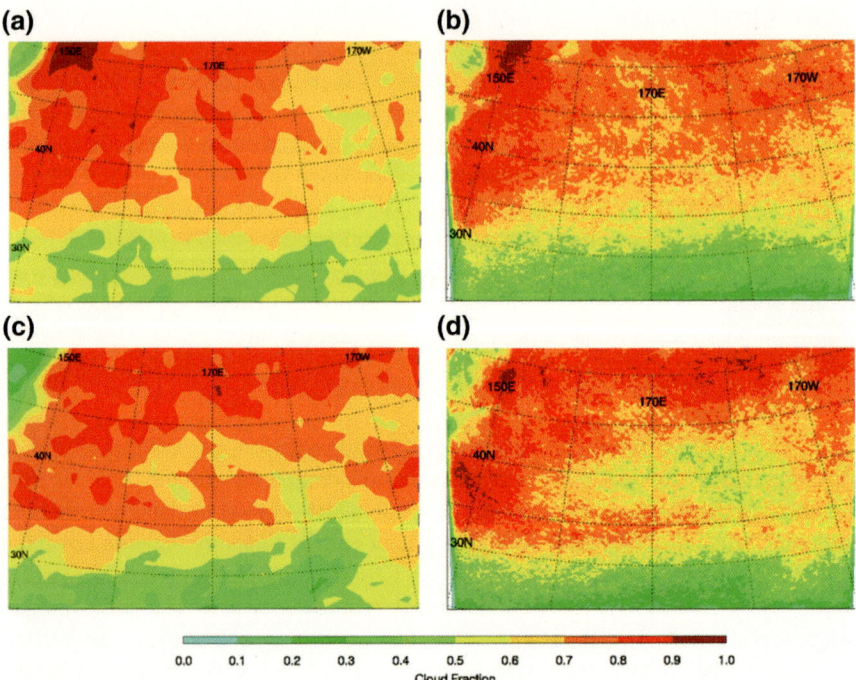

Fig. 5.7 Comparison of cloud fraction. **a** MODIS L3 monthly cloud fraction in January. **b** WRF simulated cloud fraction in January. **c** MODIS L3 monthly cloud fraction in February. **d** WRF simulated cloud fraction in February (Reprinted from Wang et al. (2014a) with permission of Nature Publication Group)

found to be greater than 50 % over the NW Pacific (Fig. 5.7). With such frequent cloud occurrences and the large cloud coverage, the atmospheric radiation is expected to be sensitive to the variation of cloud amount and albedo induced by the perturbed aerosol levels. The model simulations show that shortwave cloud forcing (SWCF) at the top of atmosphere (TOA) is increased by -1.43 W m^{-2} (Fig. 5.9a), which is corresponding to 11 % of SWCF in M-case, due to the stronger reflectance of brighter clouds with elevated aerosol concentrations. The longwave cloud forcing (LWCF) at TOA, which is dominant by high-altitude cirrus cloud or widespread anvil, provides warming effect to the radiation budget especially at nighttime. Figure 5.9b shows that the LWCF at TOA is only increased by 0.34 W m^{-2} under the P-case, so its magnitude is less than SWCF. Hence the net cloud radiative forcing is increased by -1.09 W m^{-2} in comparison of the two cases (Fig. 5.9c). The 20 % relative increase of the radiative forcing indicates that the overall cooling effect imposed by clouds onto the atmosphere is enlarged by elevation of aerosol levels over the NW Pacific region.

A belt of intensive rainfall associated with precipitating frontal cyclones is the major characteristics of the storm track. Some of the previous studies suggested that, on the global scale, the anomalous net radiative cooling induced by aerosols

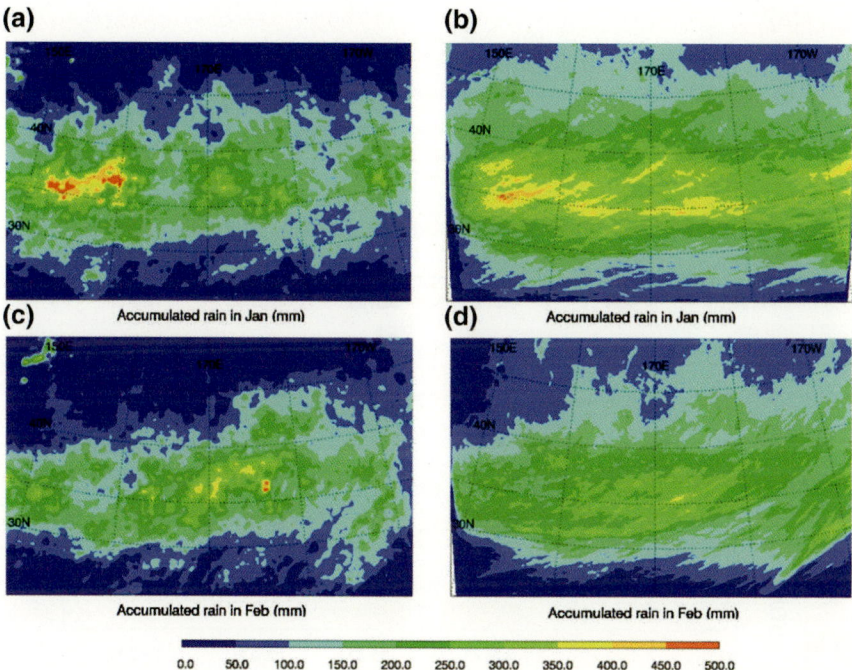

Fig. 5.8 Comparison of precipitation. **a** TRMM monthly accumulated rainfall in January. **b** WRF simulated monthly accumulated rainfall in January. **c** TRMM monthly accumulated rainfall in February. **d** WRF simulated monthly accumulated rainfall in February

was associated with reduction of the turbulent transfer of both sensible and latent heat from the surface, which finally led to weakening of the precipitation and intensity of hydrological cycle (Ramanathan et al. 2001). However, over the NW Pacific, the model simulation conducted by Li et al. (2008a) suggested 11 % enhancement of precipitation for a single storm event under the high aerosol loading through aerosol-microphysics-dynamics feedback. Therefore, it is valuable to examine these two competing effects of aerosols on the precipitation during a season period and the long-term responses of the convective cloud systems over the NW Pacific region to the different aerosol conditions. The spatial distributions of the 2-month averaged precipitation over the NW Pacific are compared between the different aerosol conditions in Fig. 5.10. Although the relative change of precipitation is not uniform over the domain, the enhancement of precipitation in the P-case can be found over the precipitation belt along the storm track. The temporal evolutions of daily precipitation for the two cases reveal that the enhancement of precipitation is about 7 % under the influence of Asian pollution outflow and consistent throughout the whole simulated winter season. The positive sign of precipitation change under high aerosol loading supports that the aerosol invigoration effect on deep convection and heavy precipitation is predominant in the wintertime NW Pacific region compared to the inhibition effect of aerosols on precipitation under the feedback from increased cloud albedo.

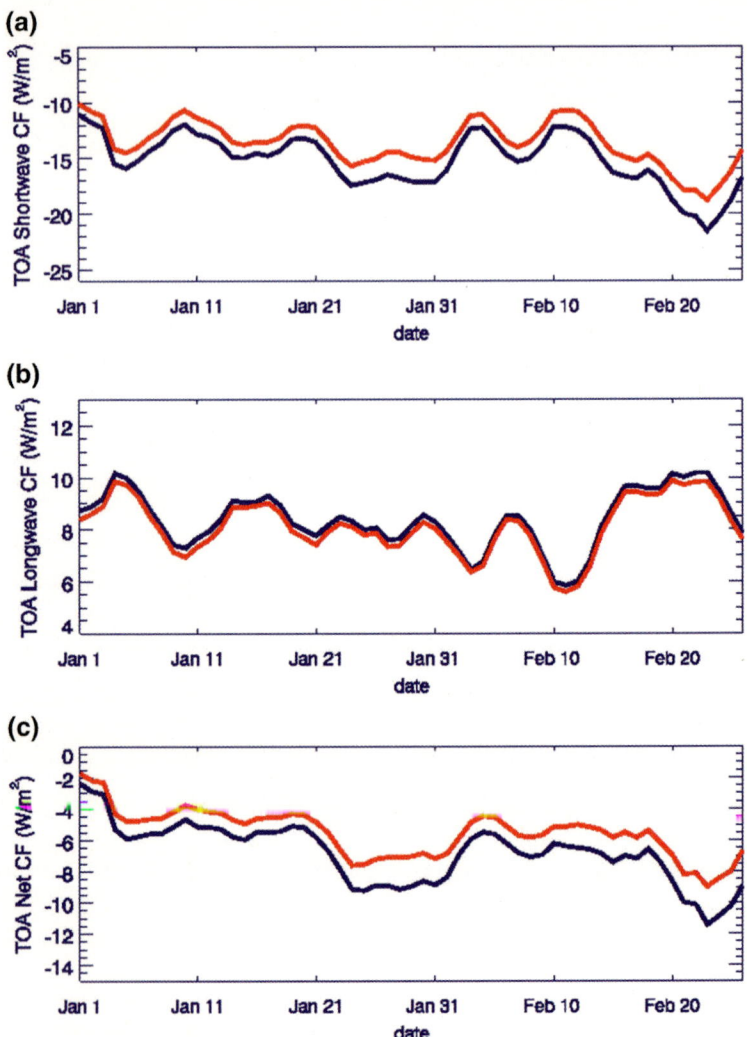

Fig. 5.9 Temporal evolution of domain-averaged TOA **a** shortwave forcing. **b** Longwave forcing and **c** net forcing. 5-day smoothing is employed. *Dark blue lines* represent P-case and *red lines* represent M-case

The physical mechanism of aerosol invigoration effect is supported by the evaluation of long-term variation of cloud properties over the storm track region. The temporal evolutions of cloud properties during the 2 months (Fig. 5.11) reveal that in the case considering Asian pollution outflow, the cloud number concentration is increased largely due to efficient activation of CCN, but the cloud effective radius is reduced by 40 %. Those results are in agreement with many in situ aircraft measurements of cloud water to rain water and the prolonged water vapor

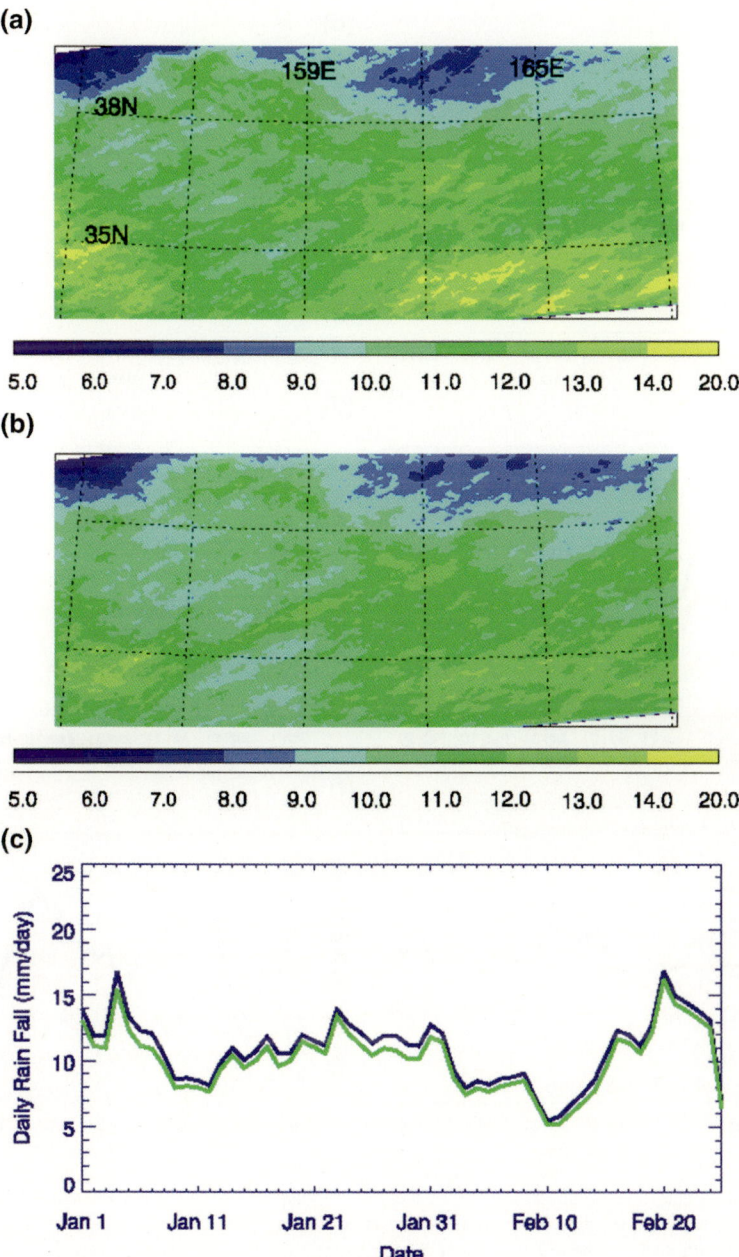

Fig. 5.10 Spatial distribution of 2-month averaged daily precipitation **a** in P-case and **b** in M-case. **c** Temporal evolution of domain-averaged precipitation rate. 5-day smoothing is employed. *Dark blue lines* represent P-case and *red lines* represent M-case

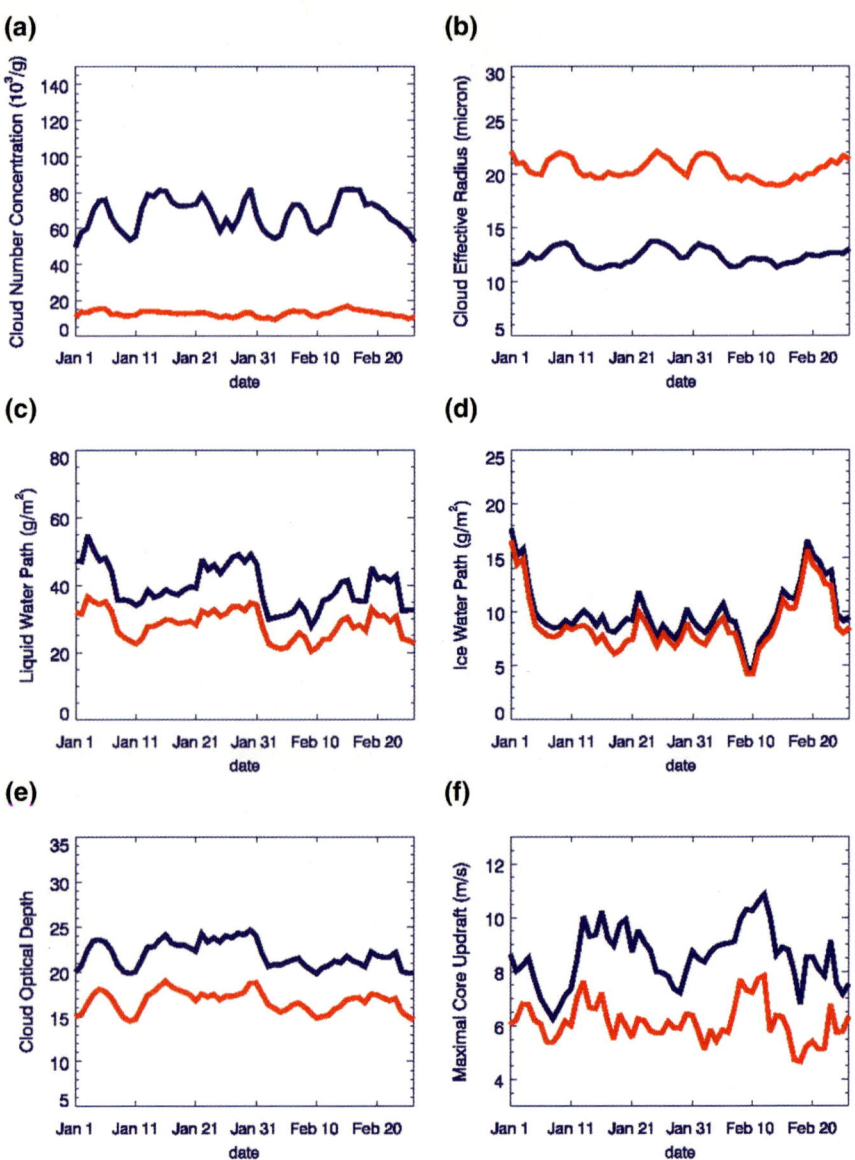

Fig. 5.11 Time series of domain-averaged **a** cloud number concentration, **b** cloud effective radius, **c** cloud water path, **d** cloud ice path, **e** cloud optical thickness and **f** cloud core area percentage. 5-day smoothing is employed. *Dark blue lines* represent P-case and *red lines* represent M-case

condensation processes lead to satellite observation under the maritime conditions (Pawlowska and Brenguier 2000). Under the influence of size reduction for cloud droplets, suppression of the warm rain processes and the prolonged water vapor condensation processes lead to the increase of cloud liquid water path by 43 %

(Fig. 5.11c). With the strong vertical motion of subtropical cyclones, averaged 12 % increase of ice water path (Fig. 5.11d) during the 2-month simulation support that more small cloud droplets are lifted to the mixed-phase or even higher levels, and more liquid water is available for homogeneous and heterogeneous ice nucleation in the P-case. The enhanced liquid and ice water contents contribute to the intensified release of latent heat from more efficient condensation, deposition, and freezing processes. The examination of vertical convection strength, which is quantified by the maximal updraft velocity over the core area (updraft velocity larger than 1.0 m/s and cloud water content larger than 10^{-5} kg/kg) (Li et al. 2008b), reveals that the maximal updraft is increased by 40 % in the P-case (Fig. 5.11f) and the dynamics of mid-latitude cyclones have been changed by introduced aerosol particles.

To quantify the comprehensive aerosol forcings from CR-WRF simulations, the total diabatic heating rate are derived from the 2-month results in the cloud-resolving domain. As shown in Fig. 5.12, the diabatic heating in WRF is composed of the summation of longwave radiation heating, shortwave radiation heating, heating from planetary boundary layer (PBL) scheme, and heating from microphysics (latent heating). Note that heating from cumulus scheme is not considered since no cumulus scheme is used in the WRF simulations.

From the profile of radiative heating between P-case and M-case in Fig. 5.12a, b, we find that the sensitivities of radiative heating to the aerosol perturbation is very limited, since the heating rate is averaged over all the grid points of the domain and aerosol effect on the clear-sky radiation is not considered in this study. In the boundary layer, reductions of the longwave radiative cooling and the shortwave radiative heating occur due to the higher cloud fraction in the polluted case, which is consistent with the discussed aerosol effects on the radiation budget in Fig. 5.9. The net effect of aerosol from the radiation scheme (Fig. 5.12c) is the net heating within the boundary layer. The heating profiles from PBL scheme (Fig. 5.12d) indicate that more heat is redistributed to the upper PBL due to the enhanced vertical sub-grid flux under the polluted case, while significant cooling occurs close to model bottom levels. The responses of latent heating to the different aerosol scenarios are complicated. We separately examine the heating from latent heat release due to condensation, deposition, freezing, and the cooling from some other phase changes including evaporation, sublimation, and melting. We notice that the P-case predicts both stronger heating and stronger cooling from the microphysics in the polluted case. Because of the delayed warm rain processes and the enhanced mixed-phase processes, larger liquid/ice water content lead to more efficient latent heat release. Meanwhile, the ice hydrometers above the freezing level in the polluted scenario will undergo melting and sublimation during the settling process which explains the peak value at about 800 mb in the cooling rate profile. Those results are consistent with the finding of aerosols impacts on convective heating by Khain et al. (2005). The overall effect is that there is net heating in the PBL and middle troposphere and net cooling around the freezing level in Fig. 5.12e. Combining all those heating terms, the mean differences of total diabatic heating rates between P-case and M-case are presented in Fig. 5.12f. The diabatic heating is found to maintain the mean maximum in baroclinicity in the storm track region (Hoskins and Valdes 1990). Therefore, in the next

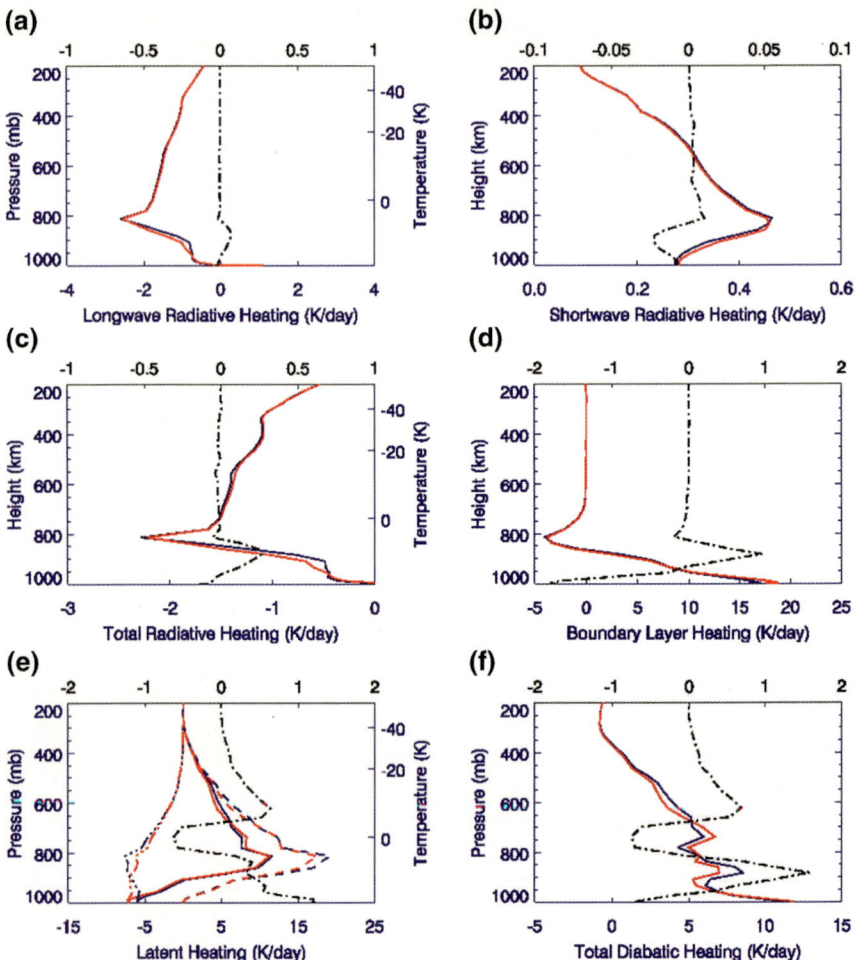

Fig. 5.12 Heating rate profiles from CR-WRF simulations. The *blue lines* denote the heating rates from P-case and red lines for M-case. The *black dot-dash lines* denote the heating difference between P-case and M-case. In (**e**), the *dash lines* denote the latent heat release rates (positive only) and the *dot-dot-dash lines* denote the cooling (negative only) rates

step, we implement this heating rate anomalies induced by the elevation of aerosol concentration into the Northwest Pacific region in the GCM and investigate the response of large-scale dynamics to the additional heating.

5.2.4 Response of Storm Track in the Forced CAM5

The stand-alone NCAR Community Atmospheric Model version 5 is employed in this study and the prescribed additional heating profile from CR-WRF is implemented in the physics package of CAM5 following the similar methodology used by Lappen and Schumacher (2012). The year 2000 present-day climatological scenario is adopted in simulations. The sea surface temperature and sea ice are prescribed in the sensitivity study to exclude the interannual variation of those dynamical factors. The model runs at $0.9° \times 1.25°$ horizontal resolutions with 30 vertical levels. Two sets of simulation are designed. The control simulations (CTRL) are conducted from the beginning of October to the end of February next year. First 3 months are used for model spin-up. In the simulations with the consideration of aerosol forcing (AERO), heating profiles will be added to the storm active region over the NW Pacific (150°–180° E, 30°–40° N, as shown in Fig. 5.13) from January to February. Three winter seasons are simulated with the random perturbations on the initial condition. The results discussed below are averaged over the three winters.

We examine the large-scale response of the storm track to the additional heating through the comparison of the transient eddy meridional heat flux at 850 hPa (EMHF) and transient eddy meridional wind variance at 300 hPa (EMHF) between CTRL and AERO. As shown in Fig. 5.14, both two types of simulation reproduce the bell-shape region with large positive EMHF and EMWV along the storm track downstream development. However, the differences between CTRL and AERO indicate that the intensity of the storm track are significantly enhanced under the influence of the modulated diabatic heating in the AERO. The location and magnitude of the EMHF and EMWV enhancement over the NW Pacific are well consistent with those of decadal differences from NCEP and ECMWF reanalysis datasets shown in Figs. 5.2 and 5.3. Moreover, the better agreement of EMHF and EMWV between AERO and 2002–2011 Reanalysis mean data demonstrates the validity of our hierarchical modeling approach adopted in this study.

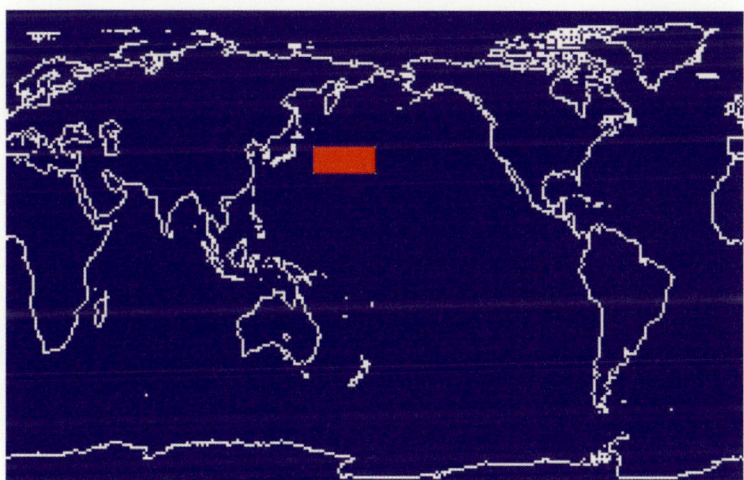

Fig. 5.13 Location of additional heating in the GCM domain

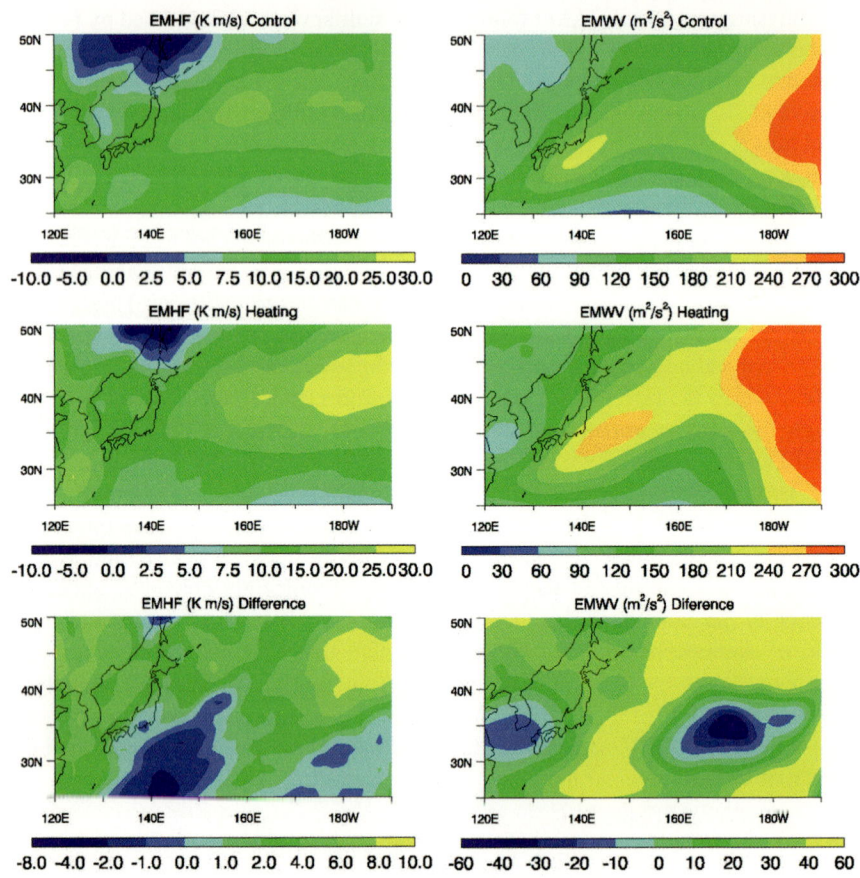

Fig. 5.14 Comparison of EMHF and EMWV between CTRL and AERO

To illustrate the physical mechanism behind the enhanced storm track intensity with aerosol forcings, we examine the vertical profiles of temperature averaged over the NW Pacific from CTRL and AERO. The temperature differences between CTRL and AERO reveal that the heating anomalies mainly occur in the middle troposphere, while there is a significant cooling at 200 mb level. Thus, the large positive buoyancy is produced in the middle troposphere of AERO simulations, which contributes to a larger baroclinic instability (Hoskins and Valdes 1990) and is favorable for the storm track dynamics. Since the additional heating induced by aerosols is close to the boundary layer (Fig. 5.12), the larger temperature anomalies in the middle troposphere in AERO are likely caused by the intensified upward transport of heat. To validate this hypothesis, we further investigate the convection strength between CTRL and AERO (Fig. 5.15).

From Fig. 5.16, it is clear that maximum updraft velocities in AERO are higher than that in CTRL at different altitudes, especially in the boundary layer and the

Fig. 5.15 Comparison of temperature profiles between CTRL and AERO. *Dark blue line* denotes the profile from AERO, *red line* denotes the profile from CTRL, and the *dash line* denotes the difference between AERO and CTRL

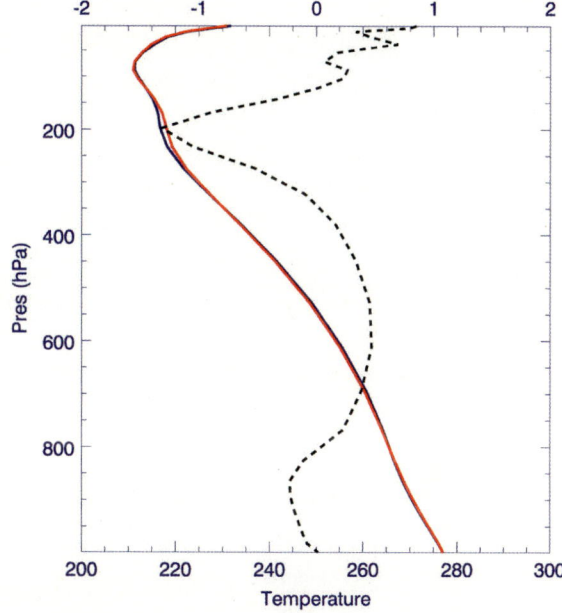

Fig. 5.16 Comparison of maximum updraft velocities between CTRL and AERO. The *blue line* denotes the result from AERO and the *red line* denotes that from CTRL

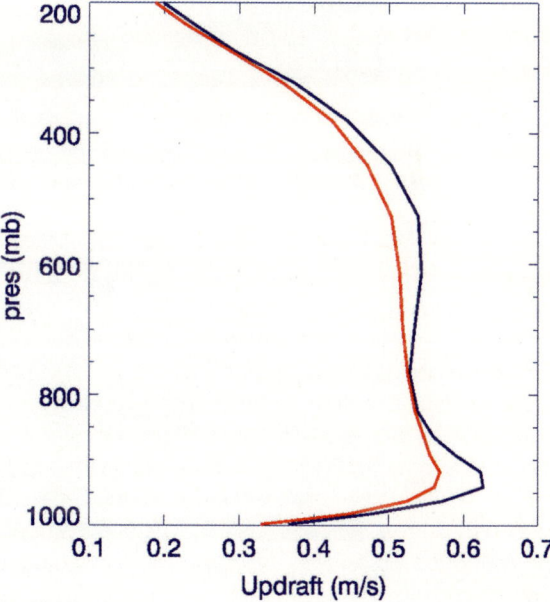

middle troposphere, which indicates that the convection is invigorated in AERO due to the added heating profile. Meanwhile, the comparison of the fraction of high cloud (with altitude higher than 400 mb) between CTRL and AERO in Fig. 5.17

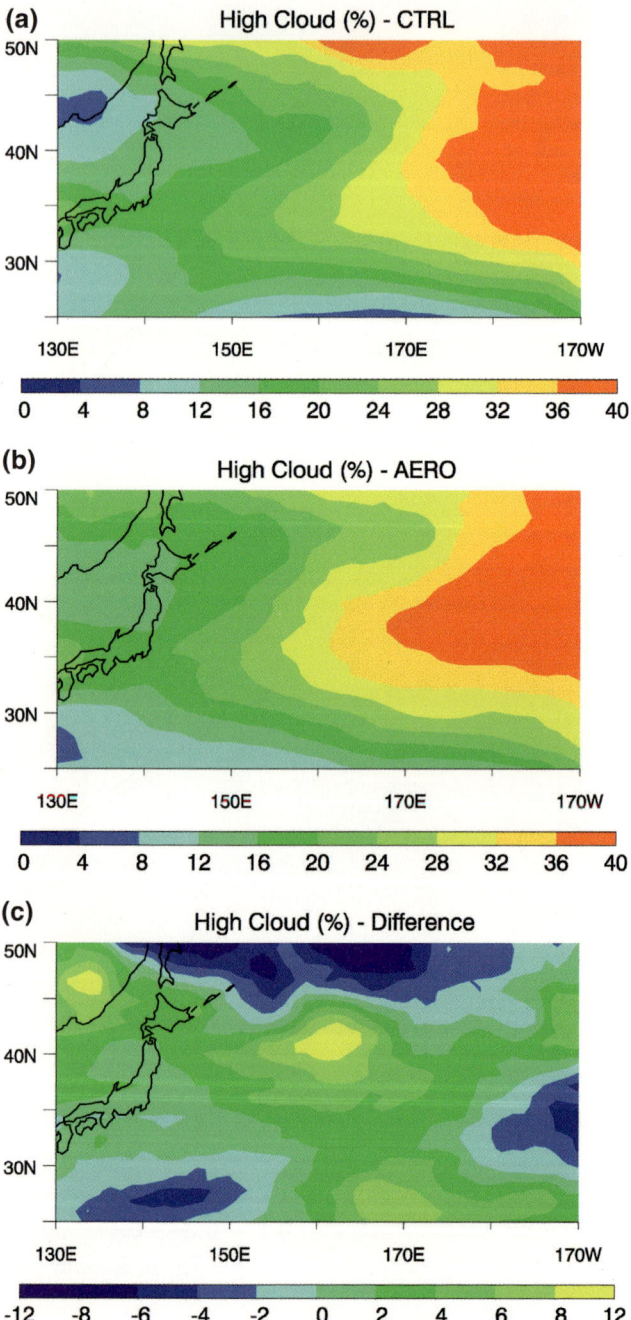

Fig. 5.17 Comparison of high cloud fractions between CTRL and AERO (Reprinted from Wang et al. (2014a) with permission of Nature Publication Group)

reveals that over the region close to the heating source (150°–180° E), larger amount of ice clouds occur in the upper troposphere under the intensified updraft velocity in AERO. This finding is also consistent with the analysis of ISCCP satellite products about decadal trend of deep convective clouds under the influence of Asian pollutions by Zhang et al. (2007).

5.3 Multiscale Aerosol-Climate Modeling Framework

5.3.1 Numerical Experiment Design

Six-year global climate simulations driven by the climatological sea surface temperature (SST) were carried on using PNNL-MMF. Simulations with the conventional CAM5 alone were conducted as well to evaluate the performance of PNNL-MMF. Both the host of MMF and CAM5 run at $1.9° \times 2.5°$ horizontal resolution with 30 vertical levels. The years 2000 and 1850 are chosen to represent the present-day (PD) and the pre-industrial (PI) time, respectively. Anthropogenic SO_2, black carbon (BC), and primary organic matter (POM) emissions are from the IPCC AR5 emission dataset. Emissions of other precursors and formation mechanisms of secondary aerosols are described by Wang et al. (2011a, b) in detail. In this study, we focus on the NW Pacific area (25° N–50° N, 120° E–170° W), referred to as the analysis region hereafter. Monthly simulation results in January and February are investigated, with respect to the fact that the Pacific storm track is only active at wintertime (Nakamura 2002) and the Asian pollution outflow peaks from January to March each year (Liu et al. 2003).

5.3.2 Analysis of Simulation Results

The continued increase in the emission of anthropogenic aerosols and their long-range transport are well documented by in situ and satellite measurements (Li et al. 2010). Figure 5.18a shows the spatial difference of simulated aerosol optical depth (AOD) between PD and PI over the NW Pacific. The near-shore areas closer to the Asian continent experience larger increase of AOD compared to the remote maritime areas, and the consequent zonal extension of AOD enhancement just reflects the transportation process of particulate pollution from East Asia to West Pacific. Averaged over the analysis region, AOD is elevated by 0.03 in PD case, which accounts for 46 % relative increase defined as the ratio of mean difference between PD and PI over the mean value of PI. The analysis of the simulated aerosol mass composition of the different chemical species in the accumulation mode over the NW Pacific reveals that the mass mixing ratios of anthropogenic aerosols, such as sulfate and black carbon, are found to be enlarged by 3–5 times in PD compared

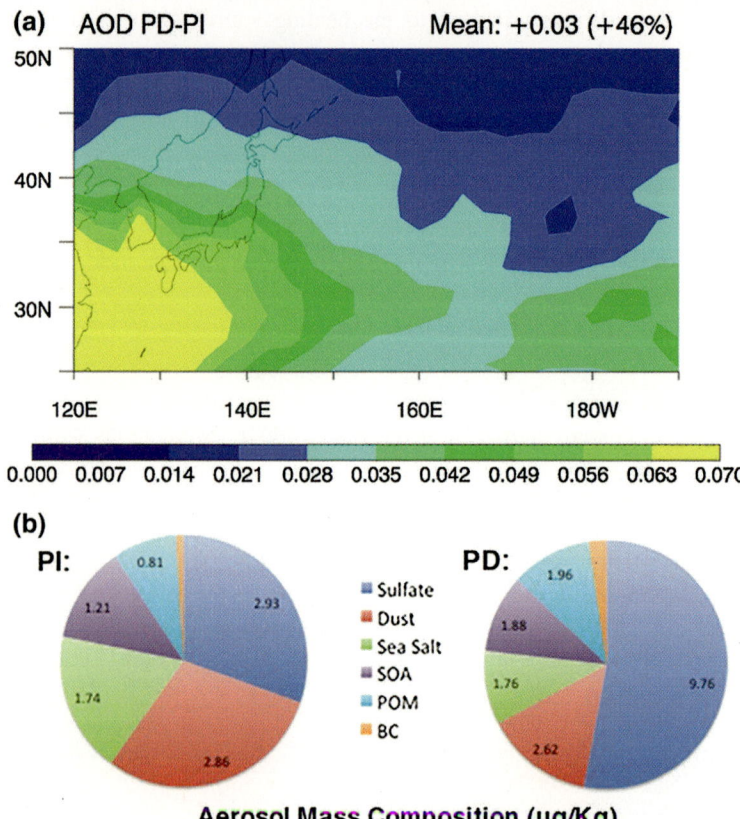

Fig. 5.18 a The difference of aerosol optical depth (AOD) between PD and PI over NW Pacific. **b** The comparison of aerosol mass concentration and chemical composition in the accumulation mode between PD and PI over the NW Pacific in PNNL-MMF (Reprinted from Wang et al. (2014b) with permission of the National Academy of Sciences of the United States of America)

to PI. The ratio of sulfate out of the total aerosol mass concentration increases from 30 to 53 %. In contrast, the levels of the aerosols from natural sources, such as dust and sea salt, are relatively less changed at the two different scenarios. Hence the enhancement of AOD over the NW Pacific in the PD scenario is mainly attributed to the elevation of anthropogenic aerosol loading in the Asian continent.

The extra-tropical migrating cyclones are prevalent in the wintertime North Pacific. The influence from the coupling between convective clouds and Asian pollution outflows on the cloud properties, atmospheric radiative forcing, and poleward heat transport are examined in Fig. 5.19. Similar to spatial pattern of AOD over the NW Pacific, cloud droplet number concentration in the costal and near-shore areas is significantly increased in PD (Fig. 5.19a). The cloud droplet number concentration averaged over the NW Pacific is enhanced by about 108 %, resulting from the elevation of sulfate aerosol concentration in PD. Meanwhile, the total

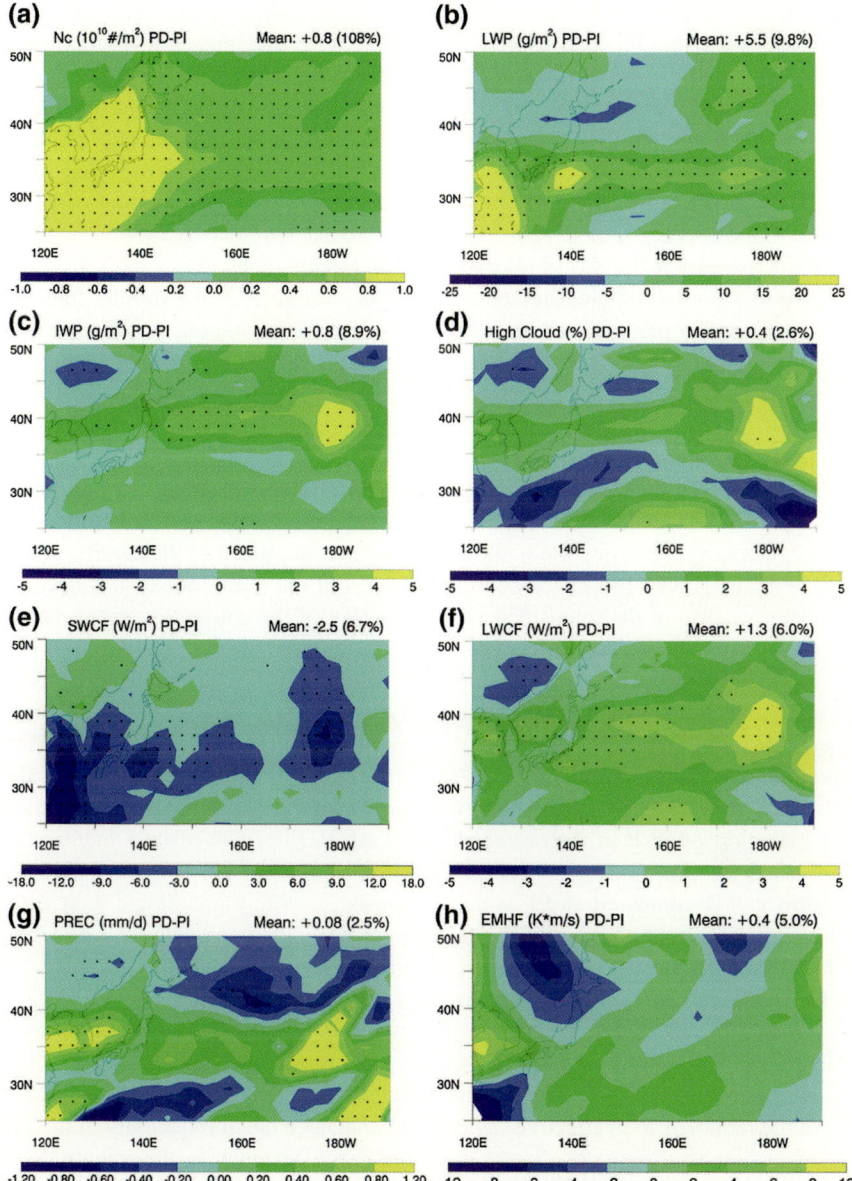

Fig. 5.19 The difference of **a** cloud number concentration (Nc), **b** liquid water path (LWP), **c** ice water path (IWP), **d** high cloud fraction, **e** shortwave cloud radiative forcing (SWCF), **f** longwave cloud radiative forcing (LWCF), **g** precipitation (PREC) and **h** eddy meridional heat flux at 850 mb (EMHF) between PD and PI over the NW Pacific in PNNL-MMF. *Black dots* indicate the regions with significance of t-test larger than 90 % (Reprinted from Wang et al. (2014b) with permission of the National Academy of Sciences of the United States of America)

liquid water path (LWP) exhibits 5.6 g/cm^2 (9.8 %) enhancement in the PD case compared with the PI case (Fig. 5.19b), indicating that the condensation gets prolonged due to the delay of warm precipitation in presence of the large amount of small cloud droplets along with the pollution outflows. Since the enhancement ratio of cloud mass content (9.8 %) is much less than the enhancement ratio of cloud number concentration (108 %) in PD, cloud effective radiuses are expected to be significantly reduced in the polluted scenario. The abundance of ice-phased particles is found to be elevated as well in the PD case. The average ice water path (IWP) increase by 0.7 g/cm^2 (9 %) over the NW Pacific in the PD case (Fig. 5.19c), suggesting the more efficient mixed-phase processes, such as the droplet freezing and riming on ice particles above the 273 K levels. As a consequence, the anvil of convective clouds may extend broader and thicker over the storm track. Note that the enhancement of the IWP is located in the center of North Pacific region (around 180° E) and close to the downstream of the storm track, different from the enhancement location of LWP that is close to the Asian continent. The examinations of the spatial distribution of LWP and IWP in PD and PI separately (shown in the supplementary material) reveal that when cyclone systems are in the embryonic stage in the west of Pacific, the liquid phase water is dominant in the clouds. When cyclone systems are fully developed in the center of the ocean, ice clouds become abundant because of the intensified vertical motion of the convection core and the followed anvil formation. The geospatial difference of the LWP and IWP enhancement over the analysis region further suggests that anthropogenic aerosols continuously exert impacts on the different types of clouds throughout the lifetime of cyclones. The fraction of high cloud above 400 mb is examined from the MMF simulations in both PD and PI cases (Fig. 5.19d). The major enhancement of the high cloud fraction is found to be close to the storm track downstream, coinciding with elevation of IWP in the center of the North Pacific. The relative increase of high cloud fraction averaged over the analysis domain is 2.6 % in the polluted scenario in PD compared to PI, which is consistent with climatologic increasing trend of the high clouds during the past a few decades reported by Zhang et al. (2007) based on the ISCCP and HIRS satellite data.

The variations of cloud radiative forcings at top of atmosphere (TOA) induced by aerosols are examined in the simulations of PNNL-MMF. The difference of shortwave cloud radiative forcing (SWCF) at TOA between PD and PI in Fig. 5.19e reveals that over the NW Pacific, the magnitude of SWCF is increased by 2.5 W/m^2 due to the reduced cloud effective size and the enhanced cloud reflectivity in the PD case, which exert stronger cooling in the atmosphere. Meanwhile, the magnitude of the longwave cloud radiative forcing (LWCF) at TOA is increased by 1.3 W/m^2 in the PD case over the analysis domain (Fig. 5.19f), providing a stronger warming effect, especially at night time. Since the spatial pattern of LWCF enhancement is highly consistent with that of the IWP and high cloud frequency, the larger amount of high-level ice clouds should be responsible for the enhanced outgoing longwave radiation at TOA in the polluted scenario. The warming effects from enlarged anvils of convective clouds induced by aerosols are consistent with the previous reports from satellite measurements (Koren et al. 2010) and CRM simulation results

(Fan et al. 2012b). The response of the surface precipitation to the different aerosol scenarios is not uniform over the NW Pacific region (Fig. 5.19g). With reference to the characteristic rainfall bell of the Pacific storm track, the enhancements of precipitation mainly occur at the areas with heavy precipitation (>6 mm/day). The precipitation rate averaged over the NW Pacific is increased about 2.4 %, implying the aerosol invigoration effect on the convective clouds during the passages of the storms. Since the mid-latitude storm track is a critical component of global general circulation to transport the heat and moisture flux, it is important to examine the meridional heat flux over the NW Pacific. The output from every 2 days from 2 year simulations of MMF is used to derive the transient eddy meridional heat flux at 850 mb (EMHF) and is smoothed by 8-day high pass filter in recognition of the typical 2–8 day period for the storms over the North Pacific. Figure 5.19h shows that the average increase of EMHF over the NW Pacific is 5 % in the polluted scenario and the increase is more significant over the storm track downstream, indicating that the heat transport associated with the Pacific storm track is intensified under the influence of the Asian pollution outflow.

The invigoration effect of aerosols on the convective storm events is further revealed through the examination of vertical profiles of convective cloud amounts and cloud top heights. Figure 5.20 shows the differences of convective cloud and cloud top fraction in vertical between PD and PI over the analysis region. Diagnosed in the MMF model by using CRM cloud statistics, the convective clouds in lower level (below 850 mb) is the slightly larger in PD than that in PI throughout the NW Pacific, while the more significant increase of the middle-level convective clouds (between 850 and 400 mb) occurs in the remote maritime area in the west of 150° E (Fig. 5.20a). The elevated cloud water contents and the strengthened updrafts are responsible for the larger amount of cloud with convective motion in the PD case. Figure 5.20b shows that the cloud top heights are dramatically changed in comparison of PD and PI scenarios. In the upper troposphere above 600 mb, the fractions of cloud top increase by 10–20 %, while clouds with heights lower than 600 mb are significantly reduced in the center of the NW Pacific from 140° E– 170° E. The modified cloud structure and elevated cloud top height provide the

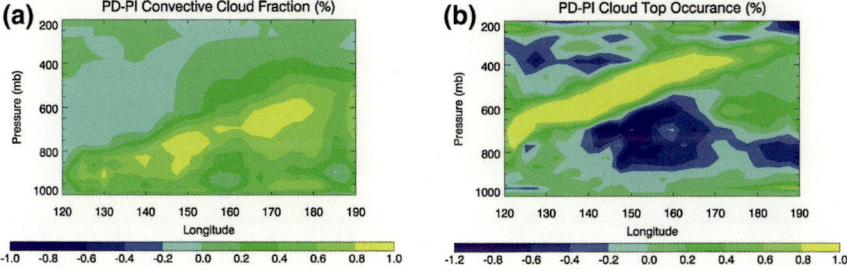

Fig. 5.20 The difference of vertical distribution of **a** convective cloud fraction and **b** cloud top fraction between PD and PI over the NW Pacific in PNNL-MMF (Reprinted from Wang et al. (2014b) with permission of the National Academy of Sciences of the United States of America)

direct evidences of the aerosol-induced invigoration effect on DCC development, which have been observed by Koren et al. (2005) over the Atlantic region using MODIS satellite data.

5.3.3 Results from Host GCM

In recognition of the ability of PNNL-MMF to reproduce the aerosol invigoration effects on the convective storms in the wintertime Pacific storm track, it is valuable to test the performance of host GCM, which cannot explicitly resolve the deep convective clouds, on simulating the aerosol-cloud-precipitation interaction in the NW Pacific. The analysis of the model simulations using CAM Version 5 is performed, and the same two emission datasets are used to represent two scenarios PD and PI for sensitivity study. As shown in Fig. 5.21, the comparison of AOD between PD and PI shows that AOD simulated in the PD case of CAM5 increases by 0.02 averaged over the NW Pacific, which has a good agreement with the 0.03 increase of AOD in PNNL-MMF. With the elevation of cloud droplet concentration and reduction of droplet effective size simulated in CAM5, LWP is enlarged by 10 g/cm^2 in PD. Thus the relative increase of LWP is 33 %, which is much higher than the 10 % relative

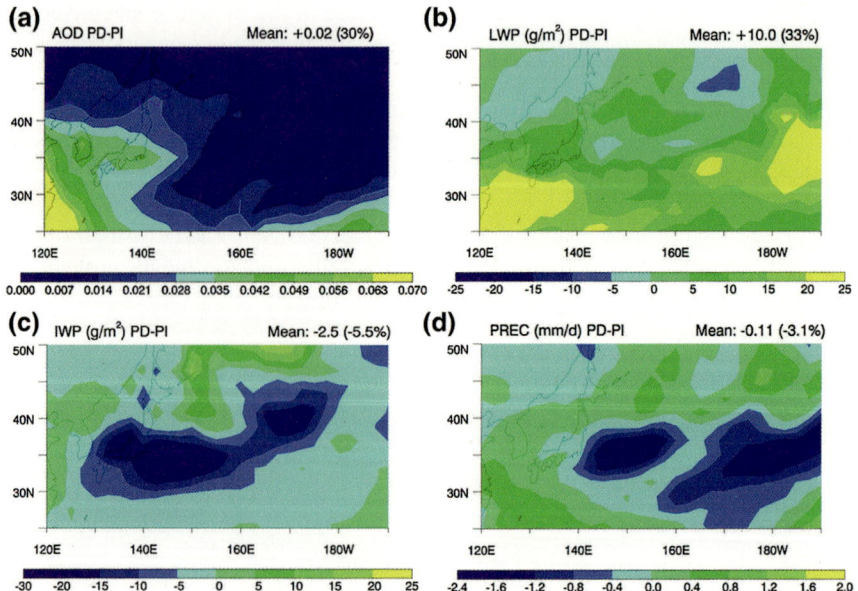

Fig. 5.21 The difference of **a** aerosol optical depth (AOD), **b** liquid water path (LWP), **c** ice water path (IWP) and **d** precipitation (PREC) between PD and PI over the NW Pacific in CAM5 (Reprinted from Wang et al. (2014b) with permission of the National Academy of Sciences of the United States of America)

increase of LWP in PNNL-MMF. It is consistent with the global comparison between PNNL-MMF and CAM5 by Wang et al. (2011b), which pointed out that the response of LWP to a given CCN perturbation in PNNL-MMF is only about one-third that in CAM5. More importantly, the IWP in CAM5 is found to be dramatically reduced over the NW Pacific region, exhibiting opposite trend to the IWP change in MMF. The convection depth and longwave cloud forcing at TOA are predicted to be reduced in the PD case of CAM simulation, indicating that the convective strength of the storms over the NW Pacific get suppressed in the CAM partially due to the absence of aerosol invigoration effect and possibly over-predicted aerosol indirect effects on stratiform clouds. The weakened storm intensity in the polluted scenario of CAM simulations further leads to the 3 % reduction of surface precipitation from the convective clouds over the NW Pacific.

5.4 Summary

The climatological variation of the Pacific storm track intensity has been examined using the NCEP reanalysis data. The eddy meridional heat flux and wind variance exhibit the strengthening trends from 1979 to 2011. The 2-month simulation using the cloud-resolving WRF model over the NW Pacific investigates the long-term effects of aerosols on the storm track region, compared to the previous simulations of an isolated storm. Driven by the same dynamic and thermodynamic forcing, sensitive experiments using CR-WRF on different aerosol loading over the NW Pacific show that, although aerosols impose a strong cooling effect by increasing the cloud albedo, there is a significant invigoration of convective cloud system development over the Pacific storm track in terms of increased precipitation, more efficient mixed-phase processes, and strengthened convection through introducing more aerosols as CCN.

To quantify the aerosol forcing associated with Asian pollution in the GCM, the diabatic heating rates between P-case and M-case in the CR-WRF simulations are examined and their differences are added to the temperature tendency equation of the CAM5 physics package. Three winter seasons are simulated with and without consideration of the additional diabatic heating over the North Pacific region. The enhanced heat flux and meridional wind variances, intensified convection, and larger amount of high clouds under the influence of modulated heating rates imply that the Asian pollution outflows exert considerable impacts onto the global general circulation with likely profound climate consequence. Therefore, the hierarchical modeling approach proves to be a promising tool to upscale the results from CR-WRF to a global climate model to better constrain the forcing terms in the global climate simulation.

Long-term global climate simulation were carried out using multiscale aerosol-climate model (PNNL-MMF) with two different precursor emissions and aerosol conditions, representing Present-Day (PD) and Pre-Industry (PI) scenarios. The comparative analysis of the PD and PI model simulations reveals that, over the NW

Pacific, the averaged cloud droplet number concentration is doubled under the influence of 46 % increase of AOD associated with Asian pollution outflows. The enhancements of cloud liquid water path and ice water path are found due to the elevated aerosol loading in the PD case, but they occur in the different locations along the storm track. The larger amount of convective cloud and the higher cloud top height in the PD case demonstrate that the convections of mid-latitude cyclones are invigorated by anthropogenic aerosols through interactions with cloud microphysics and dynamics. The greater cloud albedo and the larger amount of ice clouds lead to the enlarged shortwave and longwave cloud radiative forcing at the top of atmosphere (TOA). Surface precipitation from convective clouds and the eddy meridional heat flux with migrating storms increase by 2.4 and 5.1 %, respectively, over the NW Pacific area in PD. In contrast, CAM5 predicts the opposite signs of aerosol effects on ice water path, convection strength, and precipitation with the similar increase of aerosol loading over the NW Pacific, suggesting that the multiscale framework approach is critical for reproducing the aerosol invigoration effect on the deep convective cloud systems on the global scale.

References

Chang EKM, Lee S, Swanson KL (2002) Storm track dynamics. J Clim 15:2163–2183

Fan J, Rosenfeld D, Ding Y, Leung LR, Li Z (2012b) Potential aerosol indirect effects on atmospheric circulation and radiative forcing through deep convection. Geophys Res Letts 39(9)

Hoskins BJ, Valdes PJ (1990) On the existence of storm-tracks. J Atmos Sci 47:1854–1864

Khain A, Rosenfeld D, Pokrovsky A (2005) Aerosol impact on the dynamics and microphysics of deep convective clouds. Q J R Meteorol Soc 131(611):2639–2663

Koren I (2005) Aerosol invigoration and restructuring of Atlantic convective clouds. Geophys Res Lett 32(14)

Koren I, Remer LA, Altaratz O, Martins JV, Davidi A (2010) Aerosol-induced changes of convective cloud anvils produce strong climate warming. Atmos Chem Phys 10(10):5001–5010

Lappen C, Schumacher C (2012) Heating in the tropical atmosphere: What level of detail is critical for accurate MJO simulations in GCMs? Clim Dyn. doi:10.1007/s00382-012-1327-y

Li C, Krotkov NA, Dickerson RR, Li Z, Yang K, Chin M (2010) Transport and evolution of a pollution plume from northern China: a satellite-based case study. J Geophys Res 115 (D00K03). doi:10.1029/2009JD012245

Li G, Wang Y, Lee K-H, Diao Y, Zhang R (2008a) Increased winter precipitation over the North Pacific from 1984–1994 to 1995–2005 inferred from the Global Precipitation Climatology Project. Geophys Res Lett 35(13)

Li G, Wang Y, Zhang R (2008b) Implementation of a two-moment bulk microphysics scheme to the WRF model to investigate aerosol-cloud interaction. J Geophys Res 113(D15):D15211

Liu HY, Jacob DJ, Bey I, Yantosca RM, Duncan BN, Sachse GW (2003) Transport pathways for Asian pollution outflow over the Pacific: interannual and seasonal variations. J Geophys Res 108(D20)

Nakamura H, Izumi T, Sampe T (2002) Interannual and decadal modulations recently observed in the Pacific storm track activity and east Asian winter monsoon. J Clim 15(14):1855–1874

Pawlowska H, Brenguier J-L (2000) Microphysical properties of stratocumulus clouds during ACE-2. Tellus B 52:868–887

Ramanathan V (2001) Aerosols, climate, and the hydrological cycle. Science 294(5549):2119–2124

Roger RR, Yau MK (1989) A short course of cloud physics. Elsevier, Oxford, 293 pp

Seinfeld JH, SN Pandis (2006) Atmospheric chemistry and physics: from air pollution to climate change. Wiley, New York

Tie X, Zhang R, Brasseur G, Emmons L, Lei W (2001) Effects of lightning on reactive nitrogen and nitrogen reservoir species. J Geophys Res 106(D3):3167–3178

Trenberth KE (1997) Storm tracks in the Southern-Hemisphere. J Atmos Sci 48(19):2159–2178

Wang M et al (2011a) The multi-scale aerosol-climate model PNNL-MMF: model description and evaluation. Geoscientific Model Dev 4(1):137–168

Wang M, Ghan S, Ovchinnikov M, Liu X, Easter R, Kassianov E, Qian Y, Morrison H (2011b) Aerosol indirect effects in a multi-scale aerosol-climate model PNNL-MMF. Atmos Chem Phys 11(11):5431–5455

Wang Y, Zhang R, Saravanan R (2014a) Asian pollution climatically modulates mid-latitude cyclones following hierarchical modelling and observational analysis. Nat Commun 5:3098

Wang Y, Wang M, Zhang R, Ghan SJ, Lin Y, Hu J, Pan B, Levy M, Jiang J, Molina MJ (2014b) Assessing the effects of anthropogenic aerosols on pacific storm track using a multi-scale global climate model. Proc Natl Acad Sci USA 111(19):6894–6899

Xie PP, Arkin PA (1997) Global precipitation: A 17 year monthly analysis based on gauge observations, satellite estimates, and numerical model outputs. Bull Am Meteorol Soc 78(11):2539–2558

Zhang R, Li G, Fan J, Wu DL, Molina MJ (2007) Intensification of Pacific storm track linked to Asian pollution. Proc Natl Acad Sci USA 104(13):5295–5299

Zhao C et al (2006) Aircraft measurements of cloud droplet spectral dispersion and implications for indirect aerosol radiative forcing. Geophys Res Lett 33:L16809. doi:10.1029/2006GL026653

Chapter 6
Conclusions

The studies in this dissertation aim at advancing our scientific understandings of physical processes involved in aerosol-cloud-precipitation interactions in atmospheric science and quantitatively assessing the impacts of man-made aerosol particles on cloud systems and global circulation. A comprehensive investigation of impacts of atmospheric aerosols on different cloud systems with diverse scales over the globe has been performed on the basis of observational data and numerical simulations. The distinct ways of atmospheric aerosols interacting with cloud microphysics and the dynamics in the different cloud regimes have been revealed in this study. Meanwhile, the diverse scales of different cloud systems, ranging from urban-scale thunderstorm over North China, regional stratocumulus cloud deck over Southeast Pacific, to the large-scale Pacific storm track, pose a great challenge for model simulations. A hierarchical modeling approach is introduced to upscale the cloud-resolving model which results in global simulations to provide a realistic constraint from aerosol forcings. Simulation results from multi-scale aerosol-climate modeling frame are adopted to validate the interaction between aerosols and large-scale dynamics. Microphysics parameterizations in the regional and climate model have been evaluated as well for accurate assessment of aerosol indirect effects.

In the polluted urban area, that the occurrence of heavy rainfall (>25 mm per day) and frequency of lightning strikes are reversely correlated to local visibility cloud-resolving model simulations suggest that precipitation and lightning potential associated with thunderstorm events are enhanced by about 16 and 50 %, respectively, under the polluted aerosol condition. Moreover, elevated aerosol loading suppresses light and moderate precipitation (less than 25 mm per day), but enhances heavy precipitation. The responses of hydrometeors and latent heat release to different aerosol loadings reveal the physical mechanism for precipitation and lightning enhancement in the megacity area, showing more efficient mixed phase processes and intensified convection under polluted aerosol condition.

For maritime warm stratocumulus clouds (Sc) over the southeast Pacific Ocean, aerosols are found to be critical for cloud formation, optical properties, as well as drizzle precipitation. In the WRF model, the prescribed aerosol approach produces unreliable aerosol and cloud properties throughout the simulation period, compared against atmospheric observations and simulations produced by a spectral bin

© Springer-Verlag Berlin Heidelberg 2015
Y. Wang, *Aerosol-Cloud Interactions from Urban, Regional, to Global Scales*,
Springer Theses, DOI 10.1007/978-3-662-47175-3_6

microphysical scheme, although all of the model simulations are initiated by the same initial aerosol concentration on the basis of field observations. The prognostic double-moment aerosol representation to predict both aerosol number concentration and mass mixing ratio exhibits significant improvement on the simulation of the stratocumulus cloud properties. Sensitivity experiments using four different types of autoconversion schemes reveal that autoconversion parameterization is crucial in determining the raindrop number, mass concentration, and drizzle formation for warm Sc. An embryonic raindrop size of 40 µm is determined as a more realistic setting in the autoconversion parameterization. The saturation adjustment employed in calculating condensation and evaporation in the bulk scheme is identified as the main factor responsible for large discrepancies in predicting cloud water in the Sc case, suggesting that an explicit calculation of diffusion growth with predicted supersaturation is necessary to improve the bulk microphysics scheme. Lastly, a larger rain evaporation rate below clouds is found in the bulk scheme in comparison to SBM simulation, which may contribute to a lower surface precipitation in the bulk scheme.

Over the Northwest Pacific, the intensity of the wintertime storm track has been found to increase since 1979 on the basis of two Reanalysis datasets. To explore the possible linkage to the Asian pollution outflows, seasonal (2-month) simulations using a CR-WRF model with a two-moment bulk microphysics have been performed. Analysis over the cloud-resoving scale indicate the aerosol invigoration effect on the Pacific storm track in terms of enhanced cloud amount, increased precipitation, and intensified convection. Subsequently, the anomalies of diabatic heating rate induced by the elevation of aerosols from the Asian continent in CR-WRF simulations have been prescribed in the NACR Community Atmosphere Model (CAM5) to provide the aerosol forcing terms. The forced GCM has well reproduced large-scale responses of storm track intensity to aerosol forcings in comparison with unforced model simulations.

The analyses of two aerosol scenarios corresponding to present-day and pre-industrial conditions from multi-scale global climate model simulations also indicate significant changes in the aerosol optical depth, cloud number concentration, and cloud and ice water paths over the Northwest Pacific region. A higher cloud albedo and a larger amount of ice clouds with elevated aerosol loading lead to increased shortwave and longwave cloud radiative forcing at the top of the atmosphere by 2.5 and 1.3 W/m^2, respectively. The increased precipitation and poleward heat transport for the present day case reveal invigorated mid-latitude cyclones. Note that this is the first time that the aerosol invigoration effect on deep convective clouds is reported from the free runs of GCM. Therefore, our results suggest that the multi-scale framework approach is critical in evaluating the aerosol invigoration effect on deep convective cloud systems on the global scale.

Printed by Printforce, the Netherlands